ADVANCED
COMMON CORE
MATH
EXPLORATIONS

ADVANCED COMMON CORE MATH EXPLORATIONS

GRADES 5–8

Probability & Statistics

JERRY BURKHART

Routledge
Taylor & Francis Group

NEW YORK AND LONDON

First published in 2017 by Prufrock Press Inc.

Published 2021 by Routledge
605 Third Avenue, New York, NY 10017
2 Park Square, Milton Park, Abingdon, Oxon OX14 4RN

Routledge is an imprint of the Taylor & Francis Group, an informa business

Copyright © 2017 by Taylor & Francis Group

Cover design by Raquel Trevino and layout design by Allegra Denbo

ISBN 13: 978-1-0321-4434-4 (hbk)
ISBN 13: 978-1-6182-1546-8 (pbk)

DOI: 10.4324/9781003232780

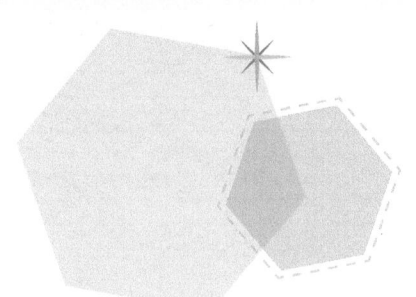

Table of Contents

Preface. vii
A Note to Students. ix
A Note to Teachers. xi
Introduction 1
Connections to the Common Core State Standards. 3
Teacher's Guide. 5

Exploration 1: Playing With Data. 17

Exploration 2: A Day at the Races. 41

Exploration 3: Simulation Station. 67

Exploration 4: Comparing Populations. 89

Exploration 5: One More Time!. 117

Exploration 6: What Are the Chances? 147

Exploration 7: Paths and Pascal 175

Exploration 8: Sports Correlations. 195

Exploration 9: Triangle Trials 233

Appendix. 251
References 255
About the Author 257
Common Core State Standards Alignment 259

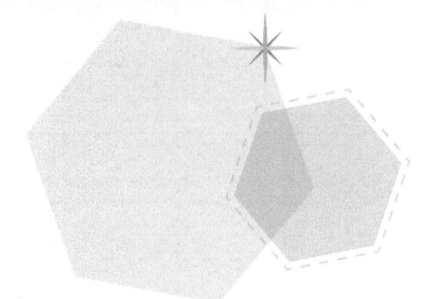

Preface

Many people have played key roles in the books in this series. I would first like to thank Sarah Scott, without whose passion, support, and imagination the books would not exist. My daughter, Annie, has greatly improved my writing through many hours of careful reading, insightful suggestions, and lively discussion. Sue Wygant has offered numerous ideas to improve the usability of the activities and to help them better support the inquiry-based approach that they represent.

If you have purchased other books in this series, you may notice some changes. Ian Byrd of byrdseed.tv has provided wonderful feedback on the format, resulting in a more appealing, open appearance and a more informal, streamlined writing style. I am grateful for his willingness to share his expertise, and I am excited for the potential to make challenging mathematical content more broadly accessible.

I would also like to acknowledge the staff at Prufrock Press. I appreciate their investment in and support of such a large undertaking. Lacy Compton has been wonderfully patient and skilled in balancing the requirements of her job as editor with respecting my vision and purpose as an author.

Finally, I would like to thank my students over the years. Their love of learning has been an inspiration to me, and these books are full to the brim with their ideas!

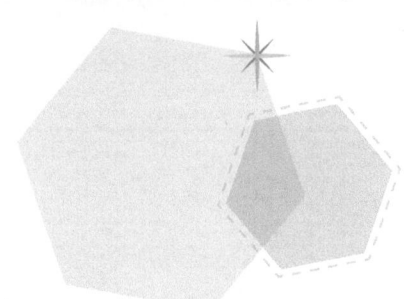

A Note to Students

Welcome, math explorers! You are about to embark on an adventure in learning. As you navigate the mathematical terrain in these activities, you will discover that "doing the math" means much more than calculating quickly and accurately. It means using your creativity and insight to question, investigate, describe, analyze, predict, and prove. It means venturing into unfamiliar territory, taking risks and finding a way forward when you are not sure which direction to go. And it means making discoveries that will expand your mathematical imagination.

Of course, the job of an explorer is hard work. At times, it will take a real effort on your part to keep going. You may spend days or more pondering a single question. Sometimes, you might even get completely lost. The process can be demanding—but it is also rewarding. There is nothing quite like the feeling of making a breakthrough after a long stretch of hard work and seeing a whole new world of ideas open up before your eyes!

These explorations are challenging, so you might want to team up with a partner or two on your travels—to discuss plans and strategies and to share the rewards of your hard work. Even if you don't always reach your final destination, you will find that the journey was worth taking. So, gear up for some hard work and adventure . . . and start exploring!

A Note to Teachers

When teachers see these explorations for the first time, some believe them to be too challenging. My experience has been different. I believe that we tend to underestimate our students. It is true that they will not typically be able to complete every problem in an activity. They are unlikely to answer most questions perfectly. Certainly, they will not finish quickly. They will sometimes struggle, and they may need your support during that process. But your students *can* be successful with the explorations—and so can you!

Often, when our students struggle, especially in math, we tend to believe that our job is to fix the "problem" by explaining more clearly or in a different way. But, in fact, it is when our students are struggling with mathematical ideas that there is the greatest potential for deep learning. Our job is not so much to explain as to ensure that their struggle is productive.

Your keys to success in this endeavor are patience, curiosity, an open mind, and a certain level of trust (perhaps even a leap of faith!) in yourself, your students, and the activities. Building this type of deep learning into your teaching practice takes time. Think of it as a *process* of learning and growing along with your students. Do not take on too much at the beginning. Assign fewer problems, and listen closely as your students talk and write about their ideas. As you reflect and come to understand how they are thinking about the problems, you will learn more about the math and about how to facilitate conversations that help students become authors of their own learning.

In the end, it is the *thinking* that creates the real learning. Talented students (really, *all* students) who develop a habit of thinking hard about challenging problems and ideas grow their mathematical capacity in deep and powerful ways. The times when they hit a wall and feel that they are making no progress are probably the times they are learning the most. Time and time again, my students come to class bursting with excitement, telling me, for example, that they were riding the bus home when the answer to a question they had been thinking about for days suddenly came to them. I love it when this happens, because they are able to see that their brains were engaged with the problem even when they were not aware of it! But the breakthrough occurs only because of the work they did while they felt stuck.

As powerful as it is, this way of learning can be "messy." Ideas do not come tied up in neat packages. You and your students gradually learn to live with ambiguity as you consider multiple strategies and make connections between different ways of thinking about problems. You begin to see that deep understanding develops gradually over time as your mind knits together various strands of ideas. In fact, this tolerance for ambiguity is a chief characteristic of successful mathematicians! The presence of uncertainty and confusion as you work toward understanding is an inescapable and wonderful part of the process of doing mathematics.

That is not to say that you normally leave things in this state of uncertainty. When students are working toward specific learning goals, it is important in the end to organize, clarify, and summarize what was learned. But this happens *after* students have wrestled with the ideas and *at the level* that they are prepared to make sense of them. The "debriefing" process described in the Eight Motivation Strategies of the introduction to this book will help you make this happen.

Teachers as well as students begin these activities with different levels of comfort and confidence. Some teachers do not think of themselves as "math people." Interestingly, those who fall into this category are often more successful than others in making the explorations work. They may be more open to thinking of math in new ways and more comfortable with the idea of learning from their students. They *become* math people! This is not to say that knowledge of mathematical content is unimportant. The more deeply you understand the math you are teaching, the more effective you can be. However, as I suggested before, this learning occurs over time, and you do it best by listening to your students and reflecting on your practice. I can personally vouch for this. My own understanding of mathematics has been completely transformed by my work with elementary and middle school students.

I cannot promise that using these explorations will be easy at first. Nor can I promise that all of your students will love doing these types of activities right away. Some will thrive immediately. Many will adapt fairly quickly. But, understandably, a few will initially (and sometimes stubbornly) prefer what is familiar and comfortable.

What I can say is that, with a little faith and persistence, doing activities like these can change students' approach to math in profound and positive ways. They find themselves slowly drawn into this type of thinking and begin to miss it when it is not present! It can be a transformative experience. Students may leave your class with an entirely new understanding of mathematics as a discipline and of themselves as mathematicians. Personally, watching this transformation take place has been the greatest joy in my work. I wish the same joy for you.

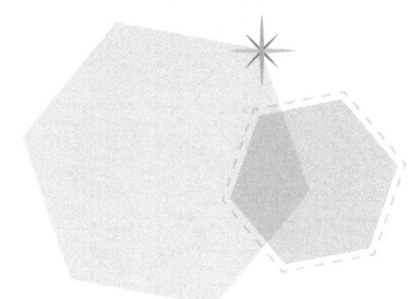

Introduction

As educators, we often take one of two approaches to math with gifted students: a serious one, in which we stick to the curriculum and accelerate to the next topic (or the next course), or a fun one, with enriching games, puzzles, and general problem solving. Both approaches can benefit students when used appropriately. However, they leave out something important.

Acceleration alone can be superficial. It may not lead to deep, meaningful learning that lasts. Enrichment by itself may be unfocused. Students develop thinking skills, but the skills are disconnected from relevant mathematical content. With a careful blend of the two approaches, you get the advantages of each and something new—a deeper, richer understanding of advanced content that prepares students for long-term success and nurtures a love of math.

The explorations in this book are both serious and fun! In some of them, the fun part jumps right out at you. But the real fun starts after you spend time thinking about the problems and begin to unravel their secrets. Motivation spurs hard work. But, with the right tasks, hard work can also create motivation!

There is a simple teaching philosophy that captures the spirit of these explorations: the students do the thinking. When we tell talented students what to do, they do it, and it's done. When we let students think for themselves, they *learn* the math. They make sense of it. They retain it, transfer it to new learning, apply it to the real world, and appreciate it. As teachers, what more could we want?

DOI: 10.4324/9781003232780-1

Connections to the Common Core State Standards

THE COMMON CORE STATE STANDARDS FOR MATHEMATICAL CONTENT

Table 1 shows the Common Core State Standards for Mathematical Content that apply to each exploration. Because the problems are rich and open-ended, the standards for each activity naturally cross grade levels and content strands. This helps students make connections between math concepts. It also makes the explorations good vehicles for differentiation.

Try to be thoughtful in interpreting the Common Core grade-level designations. The authors of the CCSSM stated, "No set of grade-specific standards can fully reflect the great variety in abilities, needs, learning rates, and achievement levels of students in any given classroom" (National Governor's Association [NGA] Center for Best Practices & Council of Chief State School Officers [CCSSO], 2010, p. 4). The learning trajectories of individuals will vary based on their experiences. Some talented students, especially those who have had many opportunities to learn through problem solving, may accelerate or even change the sequence of some learning trajectories (Johnsen, Ryser, & Assouline, 2014). Use your observations of students to make decisions about when to use an exploration or which parts of it to assign.

THE COMMON CORE STATE STANDARDS FOR MATHEMATICAL PRACTICE

The processes by which students learn and do mathematics are addressed in the eight Common Core State Standards for Mathematical Practice (NGA & CCSSO, 2010, pp. 6–8):

1. Make sense of problems and persevere in solving them.
2. Reason abstractly and quantitatively.
3. Construct viable arguments and critique the reasoning of others.
4. Model with mathematics.
5. Use appropriate tools strategically.
6. Attend to precision.
7. Look for and make use of structure.
8. Look for and express regularity in repeated reasoning.

DOI: 10.4324/9781003232780-2

TABLE 1

Common Core State Standards for Mathematical Content

Exploration	Standards
1. Playing With Data	**6.SP.A.3, 6.SP.B.5.C**
2. A Day at the Races	**6.SP.A, 6.SP.B**
3. Simulation Station	**7.SP.C** 6.RP.A.3.C
4. Comparing Populations	**7.SP.A, 7.SP.B** 6.SP.A, 6.SP.B
5. One More Time!	**7.SP.C** 6.SP.A, 6.SP.B
6. What Are the Chances?	**7.SP.C** HSS.CP.B
7. Paths and Pascal	**HSS.CP.B.9**
8. Sports Correlations	**8.SP.A.1, 8.SP.A.2, 8.SP.A.3** 8.F.B, 6.EE.C.9, 6.RP.A.3.C
9. Triangle Trials	**HSS.CP.B, HSA.REI.D**

Note: The standards shown in bold are the focus of the exploration. Other standards show connected concepts.

Johnsen and Sheffield (2013) have proposed a ninth standard to support the development of mathematical innovators: "Solve problems in novel ways and pose new mathematical questions of interest to investigate" (pp. 15–16).

The practice standards, including Johnsen and Sheffield's proposed creativity standard, are deeply embedded in the explorations. Most of these standards are present in some form in every activity. However, bringing them to life in your classroom depends at least as much upon instruction as it does upon the activities themselves.

The first step is to remember that students must be the ones doing the thinking. The Teacher's Guide in the following section provides specific support for building the mathematical practices into your instruction: teaching and motivation strategies, ideas for developing mathematical communication skills, suggestions for managing classroom conversation, and examples of implementing the explorations in different settings. The explorations themselves contain lists of observations and questions to support probing conversation, samples of multiple thinking strategies, and some actual examples of classroom discussions. Finally, the assessment tool on page 15 is targeted specifically to objectives that will keep you and your students focused on the deep, concept-oriented learning that is at the heart of the Common Core State Standards for Mathematical Practice.

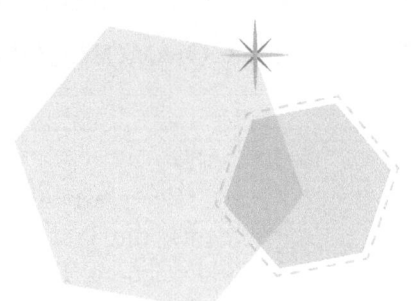

Teacher's Guide

GOALS

The explorations in this series were developed through years of work with talented middle school math students. They are designed to:

» engage students in the excitement of mathematical discovery;
» deepen students' understanding of middle school math concepts;
» help students become flexible, creative, disciplined mathematical thinkers;
» improve mathematical communication skills;
» explore connections between math concepts;
» develop patience, perseverance, and stamina in solving math problems;
» provide depth and challenge for a variety of needs and interests;
» enable students to work collaboratively and independently; and
» offer opportunities for further exploration.

THE EXPLORATIONS

This book contains problems that will challenge virtually any middle school math student. The explorations are self-differentiating. As students progress through an activity, the level of challenge or depth increases. A few students may finish. Most will reach a stopping point.

Students will progress according to their age, mathematical experience, persistence, capacity, and the amount of time available. Some may want to give up quickly. A few may insist on completing every problem even if they do not understand them well. A simple rule of thumb is that students should spend most of their time working on problems that are just beyond their comfort level. When they reach these problems, they should stick with them for a long time. They will learn more from thinking deeply about one or two problems than from rushing to finish a lot of them.

The problems in each exploration are grouped into three stages. Each successive stage extends the depth or the level of challenge. The end of a stage is a convenient place to pause and consider whether to continue. To help you decide, every stage starts with a brief description of the problems it contains along with information about the knowledge students will need and a summary of what they will learn.

5

DOI: 10.4324/9781003232780-3

Every exploration has a number of features to support your work with students: the Problems, some Conversation Starters, the Solutions, and Algebra Connections. These features are described next.

THE PROBLEM PAGE

Each Problem page has an "opener" and a list of directions. The opener is a sort of teaser that sets up the problem situation without telling students what to do. The directions fill in the details.

You may use the Problem page as a handout. My favorite approach is to cover the directions before I copy it so that students see only the opener on the handout. This is much more fun than giving them all of the information upfront. As we discuss the opener, students actively participate in creating the task by predicting (and suggesting) what the directions will be. This helps them learn that math is about asking questions, not just giving answers. It also allows me to identify possible points of confusion at the start.

At the end of this discussion, we finalize the directions. Based on students' ideas and their learning goals, we either modify the original directions or use them as is.

Some of the Problem pages have Testing the Waters or Diving Deeper questions at the bottom of the page. Testing the Waters questions are less complex versions of the main problem. They make it accessible to more students. If students are not making progress on the original problem, you can suggest that they begin with Testing the Waters. Even if they get no further, they will learn important new ideas.

The Diving Deeper questions are just what they sound like—an opportunity to explore in more depth. Many of these are more challenging than the original problem. Others point students to related questions or topics of interest.

THE CONVERSATION STARTERS PAGE

The Conversation Starters are observations and questions that can or should arise in discussion. Sometimes, your students will come up with these. At other times, you will need to work them into the conversation. Their purpose is to help you guide your students' thinking without telling them how to do the problems.

You will probably not use all of the Conversation Starters on the page. Choose those that best fit your students' needs and learning goals, or follow up on the ones that your students initiate. You do not have to use them in any particular order, but I have tried to organize them in a way that is likely to follow the flow of discussion. The Conversation Starters near the end of the page are often extensions of the main ideas.

I have written the Conversation Starters as "I wonder" questions and "I notice" statements. In keeping with the philosophy of encouraging independent thinking, many of the "I wonder" questions are not answered. I have left them open for you and your students to think about. Even when you are not sure of the answer, the question may point your thinking in a useful direction. In some cases, especially with the items near the end of the page, I raise a question out of curiosity, and I may not know the answer myself.

By the way, "I wonder" questions may pop up in the Solutions, too! You are never done with a good problem. There are always more questions to ask!

THE SOLUTIONS PAGE

In writing the Solutions, I have tried to strike a balance between giving enough detail to support your work and not giving so much that it makes the problems look harder than they are. Most solutions are one or two pages long. There are two main reasons for their length: (1) I include many student strategies, and (2) the problems contain a lot of ideas.

Please keep in mind that longer answers do not necessarily mean more classroom time. In some cases, I have simply shown many ways to think about the problem. On the other hand, a one- or two-line answer in the Solutions may represent a lot of thought and discussion. Although a solution may look short on paper, there is no shortcut for the effort and thinking that goes into finding it.

The Solutions are not the final word! You and your students may discover more efficient or more interesting strategies than I have shown. You will have insights that have not occurred to me. Each time I teach the explorations, I learn something new about the math.

THE ALGEBRA CONNECTIONS PAGE

I have written most of these explorations assuming that students have a pre-algebra level of knowledge—that they can understand, interpret, and even create algebraic expressions and equations, but they have learned few rules for manipulating them. Because your students will vary in their knowledge and experience, I have included an Algebra Connections page at the end of most activities. It has three purposes:
  » To help you see connections to students' future learning.
  » To give prealgebra students a chance to try their hand at algebraic processes and reasoning before they are taught all of the "steps" in algebra class!
  » To offer students who have studied algebra a chance to apply their skills to the problem.

If the Algebra Connections page does not seem relevant to your purposes, you may ignore it. You will not need it for other explorations. However, I hope you will glance at the connections between the content you are teaching and the concepts your students will study when you are no longer their teacher. You may gain valuable perspectives that inform your teaching. And if you feel comfortable doing so, allow your prealgebra students to play with some of the algebraic expressions and equations from time to time (without teaching them the rules)! This is a powerful way to integrate their understanding of numbers and variables.

A comment on notation: For the sake of consistency, I use a dot for multiplication and "÷" for division in most of the book. There are two exceptions. For scientific notation, I use "×" for multiplication, because it is traditional and easier to read. In the Algebra Connections, I often omit the dot and write division as a fraction. While teaching, I use a variety of notations in order to help students make the transition to traditional algebraic notation.

EIGHT MOTIVATION STRATEGIES

1. **Let students know what to expect.** Tell students that the problem or activity will take time. Let them know that they will sometimes get stuck and that their work will probably not be perfect. Give them a time frame, and let them know how you will support them.
2. **Redefine success.** Tell students that success is not just about speed and accuracy. Let them know that you value effort, progress, creativity, insight, and clear communication—in short, you care more about learning than perfection.
3. **Praise effort over ability.** Praising effort over ability encourages risk-taking. Seeing intelligence as a quantity that changes through effort empowers students to reach their potential (Dweck, 2007).
4. **Focus on process more than answers.** Respond to right and wrong answers in a similar manner, focusing on the mathematical ideas and the opportunity to learn something new. Show students that you value an interesting question as much as an accurate answer.
5. **Offer emotional support.** Some talented math students do not accept real challenges due to a fear of not looking "smart." They may not be accustomed to feeling frustration. They need help managing these feelings, especially if math has always come easily to them.
6. **Offer meaningful responses to written work.** You do not have to write a lot, just a few specific and thoughtful comments on students' completed work to let them know that you have read and considered their ideas.
7. **Allow students to collaborate.** In addition to the enjoyment of social interaction, collaboration makes students feel safer taking risks. And, of course, they have more success, because they are sharing ideas!
8. **Debrief.** After you finish a problem (or set of problems), talk about it before you move on. Kids love this! Share answers and strategies. Talk

about what went right and what went wrong. Discuss things that are still confusing. Think of new questions to ask.

TEACHING STRATEGIES

Math is about ideas! Of course, skills are necessary, too, but without a conceptual foundation, students will not be able to apply skills to problems or use them to support further learning.

Shifting from a focus on procedural skills to a more balanced approach that recognizes the key role of ideas requires thinking in new ways. The strategies in the right hand column of Table 2 show how to use these explorations to support conceptual understanding and to infuse new depth and meaning into your students' learning.

CLASSROOM DISCUSSIONS

The explorations are designed so that students may spend much of their time working without direct instruction. However, they will need to talk about the problems with you and with each other. You may need to be creative to find time for discussion, especially if you have a classroom with a wide range of needs, but it is worth the effort. The more that you and your students talk about the math, the more progress they will make and the more they will learn.

Equally important is what happens during conversation. Fortunately, you do not have to explain how to do the problems. That is your students' job! Yours is first to ensure that they have the basic knowledge needed to approach the problem and then to orchestrate conversation so that they learn from each other. The Conversation Starters give examples of questions and observations that move a discussion forward without telling students what to do. When in doubt, ask rather than answer, and say less rather than more.

To make conversations productive, classrooms must have a culture of curiosity and respect. All contributions to discussion are valuable, because all have the potential to create learning. Give students plenty of "wait time" before and after you call on them so that they have time to think and to formulate their responses. Ask them to speak in a strong voice and to direct their comments to the class rather than to you. Have them question, repeat, or rephrase each other's statements as needed. Have them agree or disagree—always explaining why. To facilitate, you may record and organize their ideas on the board. Rephrase their statements yourself for clarification if necessary, but always check that you have understood their ideas correctly. To learn more about these and other techniques for questioning and orchestrating classroom discussions, see Chapin, O'Connor, and Anderson (2013) and Smith and Stein (2011).

TABLE 2

Teaching Strategies

Traditional Strategies	Strategies That Support Deep Learning
Prepare students for guided practice by clearly explaining procedures using worked examples.	Expect students to learn by thinking their way through challenging problems that engage them with the concepts.
Teach skills first. Then have students apply them to story problems.	Use problem solving as a means of teaching concepts and skills.
Grade homework by marking answers right or wrong.	Respond to students' work by writing comments related to their thinking.
Study answers in advance so that you can explain them clearly to the students.	Be ready to discuss unexpected strategies and learn new ideas from students.
Know the process you want students to use.	Assign tasks that can be solved in many ways. Discuss advantages and disadvantages of different methods.
Have every student do the same questions.	Differentiate goals and assignments based on students' learning needs.
Have fixed deadlines for assignments.	Be flexible with due dates if students run into unexpected difficulties or want to explore further.

ASSESSING STUDENT LEARNING

The tool on p. 15 is designed to assess concept-focused tasks. It was informed and inspired by many sources: the Common Core State Standards for Mathematical Practice (NGA & CCSSO, 2010), the Process Standards of the National Council of Teachers of Mathematics (NCTM, 2000), the five Proficiency Strands in *Adding It Up* (Kilpatrick, 2001), and a rubric in *Extending the Challenge in Mathematics* (Sheffield, 2003).

You may design your own scoring system. I use a 5-point scale in each category.

- 5   evidence of learning beyond the level of course standards
- 4   evidence of learning at the level of course standards
- 3   evidence of learning approaching the level of course standards
- 2   evidence of learning below the level of course standards
- 1   evidence of learning significantly below the level of course standards
- 0   little or no evidence of progress toward meeting course standards

In my classes, students who are new to the explorations often receive 2s and 3s at first. As the school year progresses, they receive mainly 3s and 4s with an occasional 5. Students and parents appreciate the opportunity to identify specific areas of strength and goals for improvement. However, no numerical scoring system will ever replace the value of a few thoughtful written comments related to students' ideas!

Of course, you may also incorporate criteria such as legibility, organization, mechanics (spelling, punctuation, and grammar), etc. Above all, however, your scoring system should reflect the central goal of mathematical learning.

MATHEMATICAL COMMUNICATION: PART 1

Learning to communicate mathematically offers two key benefits for students. It helps them to develop their own thinking and to communicate with others. This page focuses on the first reason. When I ask young elementary students why they believe it is important to explain their thinking, they usually mention both reasons. When I ask older students, they tend to focus on the second reason. I suspect that the more they learn to think of math only as numbers and calculations, the less value they place on thinking and writing.

You have probably experienced the challenge of trying to convince some of your students to write their ideas down. They may take pride in their ability to do it all in their heads at lightning speed. This type of intuition is wonderful, but it is not always reliable, and it is not enough. When students are working on problems that are sufficiently demanding to be worth their time, there is usually too much information to manage mentally. They must write as they work in order to remember what they have done, to clarify their thoughts, to visualize relationships, to recognize and extend patterns, and to identify and correct errors. I recommend that from the beginning students have paper and pencil in front of them at all times when they are solving challenging problems. In my classes, we call it "thinking paper" rather than scratch paper in order to emphasize its purpose and its importance.

A colleague of mine who is a high school English teacher tells her students that "fuzzy writing means fuzzy thinking." I cannot think of a truer statement for math class. The effort you take to write clearly helps you to think more clearly!

MATHEMATICAL COMMUNICATION: PART 2

In order to express their ideas when solving deep and challenging problems, your students may need to expand their idea of what it means to "show work." Many important ideas cannot be captured in mathematical symbols alone. There are three common ways to communicate mathematical ideas: words, symbols (numbers, equations, etc.), and diagrams.

If students struggle with putting words on paper, suggest that they speak their ideas aloud and transfer them to paper. Of course, they will need to make some changes, but at least this gets them started. I often have students who insist that they do not know what to write—but are able to speak their thoughts perfectly clearly!

A few tips for using words:

- » Use the word "it" sparingly, and always explain what "it" is!
- » Can family or friends understand what you have written? If not, then rewrite.
- » Be concise. More is not always better.

Students may be familiar with showing their thinking using numbers and equations, but there are still a couple of pitfalls. Labels or explanations of what numbers mean are important. Also, avoid "run-on" math sentences such as $13 - 7 = 6 \cdot 2 = 12$. This statement is false, and it shows a misunderstanding of the "=" symbol. It should be written $(13 - 7) \cdot 2 = 12$ or as two separate equations: $13 - 7 = 6$ and $6 \cdot 2 = 12$.

Diagrams can sometimes communicate ideas more clearly than words or symbols. They may even replace whole sentences or paragraphs! The key is to include all of the important information without cluttering them with unnecessary or distracting detail.

EXAMPLES OF USING THE EXPLORATIONS

Example 1: Ms. Kava teaches gifted math pull-out groups of 5 to 10 students. Each group meets once per week during its regularly scheduled math time.

- » She coordinates with the classroom math teachers to select activities that align to course content.
- » Students work on explorations during the pull-out. They alternate between partner work and whole-group discussion.
- » Ms. Kava assigns "writing prompts" in which students summarize their understanding of a problem or solve a new problem from the exploration.
- » Ms. Kava shares her observations and students' completed work with the teachers.

Example 2: Mr. Hill teaches fifth-grade math in a cluster classroom with six identified high-ability math students. He has fewer students with other special needs, but he has a significant range of abilities in the classroom.

- » He uses flexible grouping. He holds two to three 15-minute math conversations with a small group of advanced students each week.
- » He uses pretests, exit slips, and informal observations to identify students for the advanced group. Cluster students participate regularly, while others flow in and out of the group as their needs indicate.

» He makes the problems available to all who are interested. He sometimes discovers students with talents he had not noticed before.

» Students work on the explorations in pairs on days he does not meet with them.

» He has other enrichment tasks not requiring instruction for students to use when they are stuck on a problem and he is not available.

» Most students work on Stage 1 of the explorations. More advanced or motivated students often continue further.

» The school is planning to implement a half hour per day for targeted instruction when students are shared between classrooms. Mr. Hill and his colleagues plan to use the time for focused instructional opportunities with the explorations.

Example 3: Ms. Rodriguez teaches a stand-alone gifted math class for sixth graders who have scored at or above the 95th percentile on a standardized math test.

» Much of the classroom instruction is based on the explorations.

» She uses her textbook primarily to sequence instruction and as a source of exercises to solidify concepts and skills.

» Many of the explorations are used as classroom lessons. Others are used as long-term (1- or 2-week) homework assignments.

» Lessons flow back and forth between small-group work on the problems and whole-group discussion in which students share and compare strategies.

» Ms. Rodriguez sets aside 10–15 minutes per day of time for class conversation about explorations that have been assigned as homework.

» Students typically complete Stages 1 and 2 of most explorations. She modifies expectations for a few students who are working hard but are having trouble finishing the homework. They complete fewer problems but are still expected to explain their thinking clearly and thoroughly. For students who need additional depth and challenge, Ms. Rodriguez makes Stage 3, Diving Deeper, or Algebra Connections tasks available as an option.

Example 4: Ms. Langford teaches the advanced section of a seventh-grade prealgebra class. Students in the class have generally scored at or above the 70th percentile on a standardized math test and had a supportive teacher recommendation. She implements a flipped classroom, in which students view daily instructional videos outside of class and complete summary tasks to verify understanding. Class time is used primarily to respond to individual needs.

» Based on her evaluation of the summary tasks, Ms. Langford creates groups whose members are in a similar place with respect to current learning goals.

» Some students spend a fair amount of group time on foundational understanding, but most are able to spend a significant portion of their time on the explorations.

» Ms. Langford takes about 10 or 15 minutes of class time to introduce a new stage of an exploration to the entire class when it comes up. Most students complete two stages of each exploration. A few complete all three. Sometimes, students skip Stage 1.

» As she monitors and assists with work on the explorations, Ms. Langford shifts students fluidly between groups in order to (1) allow those who are in similar places to work together in solving the problem, and (2) bring students into conversation when they have developed different ideas or strategies from which the other(s) can learn.

» Ms. Langford pulls groups together for a larger, focused conversation when she hears insights or misconceptions that many will benefit from discussing.

» She assigns one or two selected problems from the explorations to be written up, turned in, and graded each week.

Example 5: Mr. Okoro is the primary person responsible for teaching math at a small school. With one math class per grade level, he has a full range of needs in each class.

» He pretests students and offers the explorations to students who have demonstrated an understanding of the concepts from a lesson or unit.

» He reduces the number of practice exercises for these students and has them work in small groups on the explorations during their extra time in class.

» He meets with them as often as he can, but because he has limited time, he also makes "hint cards" from the Conversation Starters. If students have worked hard but are stuck, and he is not available to help, they take a hint card and use it to continue discussion with their peers.

» He sometimes introduces the first problem of an exploration to the whole class, using the Testing the Waters problem or other modifications to make it accessible to all. Occasionally, he discovers students who may not test well, but show signs of talent in solving nonroutine problems.

» The school has set aside a time early in the day when students meet across grade levels. He proposes using some of this time for students to receive targeted instruction and conversation time with the explorations.

Assessing Student Learning

Criterion	Description	Score
Depth of Understanding	» Know the *why* behind the *how*. » Understand the meanings of concepts. » Recognize and use connections between ideas.	
Problem Solving	» Create and use effective problem-solving strategies. » Verify your results. » Solve the problem more than one way.	
Elaboration and Communication	» Give thorough, clear, concise explanations. » Use words, calculations, and diagrams effectively. » Support your explanations with examples.	
Generalizations and Reasoning	» Recognize, analyze, and extend patterns. » Make and test predictions. » Use logic to evaluate claims and justify conclusions.	
Correctness and Precision	» Give correct answers stated with appropriate precision. » Calculate accurately and efficiently. » Use mathematical vocabulary correctly and precisely.	
Originality and Extensions	» Invent ideas and strategies that were not taught. » Find ideas and strategies that are rarely discovered. » Propose new ideas or questions to study.	
Effort and Perseverance	» Show consistent effort. » Make progress appropriate to your understanding. » Persist through difficulties.	

Exploration 1

Playing With Data

Playing With Data is the only activity in this book in which students study number relationships in data without focusing on real-world questions or making real-world decisions. (And even here they imagine a realistic context for the numbers.) However, rather than simply learn definitions and perform calculations, students create their own examples of data sets and explore *relationships* between the mean, median, range, mode, and other statistical measures.

BACKGROUND

The mean, median, and mode are measures of *central tendency*; they describe a "typical value" in a data set. The range is a measure of *variability*; it tells you something about how "spread out" the data are. In this exploration, students solve challenging problems related to central tendency and variability. They also begin to explore more precise ways to measure variability. By solving problems and thinking about relationships between quantities rather than simply doing calculations, students develop a deeper, more flexible understanding of the ideas and are better prepared to apply their knowledge to real-world situations in the later explorations.

DOI: 10.4324/9781003232780-4

STAGE 1

Before beginning Problem #1, discuss the concepts of *central tendency* and *variability* with your students. (See the Introduction to this exploration for more information.) Ensure that the students understand how to calculate the mean, median, range, and mode.

I like to show students the problem's "opener" (the part at the top of page) before I let them see the directions. I show them the four numbers and ask them what they notice and what they wonder. More often than not, they come up with the tasks in the directions on their own! The conversation gives me a better understanding of questions, misconceptions, or insights they may have. Additionally, the students become more engaged in the problem and often create their own interesting questions to explore.

What You Will Need

» Calculators

What Students Should Know

» Calculate the mean, median, range, and mode of a data set.

What Students Will Learn

» Understand the meanings of *central tendency* and *variability*.
» Explore relationships between measures of central tendency and variability.
» Solve challenging problems related to statistical measures.

Problem #1

Mean: 73.9   Median: 86.5

Mode: 26   Range: 70

Directions

- Describe a real-world situation that these numbers could represent.
- Find a set of 10 whole numbers that will produce these summary statistics.
- Justify your answer. (Discuss each summary statistic.)
- Find more solutions. Describe your thinking strategies.

Exploration 1: Playing With Data

Testing the Waters

Solve Problem #1 for five whole numbers that have a mean of 80, a median of 75, a mode of 70, and a range of 30.

CONVERSATION STARTERS FOR #1

What do you notice? What do you wonder?

I notice that the mode is much lower than both the mean and the median.

I notice that the range seems very large.

I notice that the mean is quite a bit less than the median.

I wonder what causes the mean to be less than the median?

I notice that I can predict the sum of the 10 numbers.

After students have found a solution:

I notice that I can use my first solution to find more.

I wonder if I can change the two middle numbers without changing the median?

I notice that whenever I change the two middle numbers without changing the median, the mean remains unchanged as well.

I wonder how information about the mean, median, mode, and range would be useful in the real-life situation that I created?

I wonder how many solutions this problem has?

For students who are familiar with box-and-whisker plots:

I wonder what are the smallest and largest boxes possible for box-and-whisker plots of data sets that have these same statistics?

For students who are familiar with outliers:

I wonder if the number 26 is an outlier for my data sets?

SOLUTIONS FOR #1

Sample real-world situation: The numbers could represent scores on a 100-point test.

Sample solution: 26, 26, 66, 74, 86, 87, 90, 93, 95, 96

Justification:
  » The mean is 73.9, because the sum of the 10 numbers is 739, and 739 ÷ 10 = 73.9.
  » The median is 86.5, because the two middle numbers in my ordered list are 86 and 87, and 86.5 is halfway between these numbers.
  » The mode is 26, because the number 26 appears most often in the list.
  » The range is 70, because the difference between the maximum and the minimum is 96 − 26 = 70.

More sample solutions:

$$(1) \ 26, 26, 67, 74, 86, 87, 89, 93, 95, 96$$

$$(2) \ 26, 26, 66, 73, 86, 87, 92, 93, 94, 96$$

$$(3) \ 26, 26, 66, 74, 85, 88, 90, 93, 95, 96$$

There are *many* other solutions!

Thinking strategies: Leave 26, 26, 86, 87, and 96 the same so that the median, mode, and range do not change. Increase one of the other numbers by some value, and compensate by decreasing a different number by the same amount, being careful not to repeat any values (in order not to affect the mode) and ensuring that the third and fourth numbers remain between 26 and 86 and that the seventh, eighth, and ninth numbers remain between 87 and 96. For example, increase 66 to 67 and decrease 90 to 89 as shown in solution (1). Solution (3) shows that the two middle numbers may also change as long as they maintain a sum of 173.

STAGE 2

Like many of the ideas in this book, Problem #2 came directly from questions and observations made by my students. Before sharing the directions, collect your students' ideas for new questions to ask. You may want to follow up on some of their questions.

Before students begin working independently on Problem #3, ensure that they understand the meanings of *deviation* and *absolute deviation* from the mean.

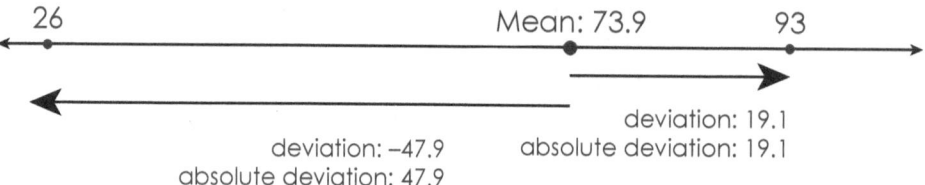

26   Mean: 73.9   93

deviation: −47.9   deviation: 19.1
absolute deviation: 47.9   absolute deviation: 19.1

Rather than simply telling students the definitions, give them a chance to use context clues and the everyday meanings of the words to predict as much as they can. The *deviation* of a data point from the mean is its distance from the mean, using negative values to account for numbers less than the mean. The *absolute deviation* is always positive, because it takes account of the distance only. (Some students may notice a connection to the term *absolute value*.)

In Problem #4, students take a closer look at sums of absolute deviations. This problem may be more appropriate for older, more advanced, or highly curious students. It (along with the Diving Deeper question) will eventually lead students to a deeper understanding of *standard deviation*, a more advanced but commonly used measure of variability.

What You Will Need

» Calculators

What Students Should Know

» Strategies and some solutions for Problem #1.

What Students Will Learn

» Analyze relationships between mean, median, range, and mode.
» Choose an appropriate measure of central tendency for a given situation.
» Explore new ways to measure variability.
» Solve challenging problems related to statistical measures.

Problem #2

Mean: 73.9   Median: 86.5
Mode: 26   Range: 70

After solving a problem, it is interesting to extend it by thinking of new questions to ask.

Directions

- Determine if the data can include the number 26 three times. Prove your answer.
- The mean, median, and mode are meant to describe a typical number in a set of data. Decide which of the three does the best job of this for your solutions. Explain your reasoning.
- Find 20 numbers that produce the same four values for the statistics. Explain your thinking.

CONVERSATION STARTERS FOR #2

What do you notice? What do you wonder?

I wonder how small the mode can be (without changing the values of the other three statistics)?

I notice that most of the numbers in my data sets are much larger than 26.

I notice that replacing one of the 10 numbers with 26 decreases their sum by quite a lot.

I notice that the mean incorporates all numbers in a data set, but the median does not.

I notice that a small change could eliminate the mode.

I wonder if I can use one of my earlier solutions to find a solution with 20 numbers?

I notice that in order to maintain the same mean, the sum of 20 numbers will have to double.

SOLUTIONS FOR #2

The possibility that the number 26 appears three times: It is not possible. So far, the data must look like this:

$$26, 26, 26, \underline{\quad}, 86, 87, \underline{\quad}, \underline{\quad}, \underline{\quad}, 96$$

Although the two middle numbers do not have to be 86 and 87, they must have a sum of 173. Therefore, changing them will not affect the sum of these six numbers, which is 347. Because all 10 numbers must have a sum of 739 (from 73.9 · 10), the four remaining numbers must have a sum of 392 (from 739 − 347).

The highest possible values for the remaining numbers are 86, 95, 95, and 96. (Because 26 appears three times, you can now use two each of 86, 95, and 96 without affecting the mode!) These have a sum of 372, which is less than the 392 that you need. Choosing different values for the two middle numbers makes things worse, because it forces the 86 to become lower, but it still does not allow the three other numbers to get higher.

Choosing an appropriate measure of central tendency: The median is probably the best choice, because the mode is much less than a typical number, and the mean is fairly strongly affected by the two 26s, making it noticeably smaller than most of the numbers in any of the solutions.

A solution with 20 numbers: There are many solutions to this problem. Some students may discover the idea of simply taking one of their previous answers and repeating each value twice! For example, based on the original solution, you would obtain:

$$26, 26, 26, 26, 66, 66, 74, 74, 86, 86, 87, 87, 90, 90, 93, 93, 95, 95, 96, 96$$

» The mean remains the same, because the sum doubles, but you are now dividing by 20 instead of 10.
» The median remains the same, because the two middle numbers stay the same.
» The mode remains the same, because 26 still appears the most often.
» The range remains the same, because the maximum and minimum do not change!

Problem #3

Mean: 73.9   Median: 86.5

Mode: 26   Range: 70

The *deviation* of the number 26 from the mean is –47.9.

Directions

- Choose one of your solutions. Find the *deviation* of each number from the mean.
- Without calculating, predict the sum of the deviations. Test your prediction.
- Repeat for at least one of your other solutions. Describe what happens and why.
- Find the sum of the *absolute deviations* from the mean for at least two data sets.
- Compare the sums of the absolute deviations of your data sets, and explain what they tell you about each set.

CONVERSATION STARTERS FOR #3

What do you notice? What do you wonder?

I notice that the deviation from the mean shows the distance from the mean.

I wonder why the deviation shown in the problem (−47.9) is a negative number?
The deviation is negative because the number is less than the mean.

After students have found the sums of the deviations:

I wonder how I can be sure that the sums of the deviations will always be 0?

I notice that absolute deviations are helpful, because their sums are not 0 (unless every number in the data set is the same, of course).

I wonder if people ever use the *mean* of the absolute deviations instead of the sum?
Yes. The mean of the absolute deviations is abbreviated MAD. See the Solutions for more information.

I wonder if data sets with smaller ranges always have smaller absolute deviation sums?

I wonder if people ever find deviations from numbers other than the mean?

I wonder if sums of deviations from numbers other than the mean would usually be greater or less than sums of deviations from the mean?

SOLUTIONS FOR #3

Deviations from the mean of all numbers in a sample solution:
- » The solution: 26, 26, 66, 74, 86, 87, 90, 93, 95, 96
- » The deviations: −47.9, −47.9, −7.9, 0.1, 12.1, 13.1, 16.1, 19.1, 21.1, 22.1
- » Prediction for the sum: Students' answers will vary. Some may predict 0.
- » The sum of the deviations: 0

Deviations for another solution:
- » The solution: 26, 26, 66, 74, 85, 88, 90, 93, 95, 96
- » The deviations: −47.9, −47.9, −7.9, 0.1, 11.1, 14.1, 16.1, 19.1, 21.1, 22.1
- » The sum of the deviations: 0

Why the sum of the deviations from the mean is equal to 0:
  Students' responses will vary. Possibilities include:
- » Once you see the sum for one solution is 0, you can tell that the others must be 0 as well, because any compensation that keeps the sum of the numbers equal to 739 will also keep the sum of the deviations equal to 0.
- » Because the mean is a "middle" number, it makes sense that values below the mean (with negative deviations) compensate for the values above the mean (positive deviations).

Some students may be able to give more precise justifications. See the Algebra Connections at the end of this activity.

The absolute deviation for the first solution:
- » The solution: 26, 26, 66, 74, 86, 87, 90, 93, 95, 96
- » Absolute deviations: 47.9, 47.9, 7.9, 0.1, 12.1, 13.1, 16.1, 19.1, 21.1, 22.1
- » Sum of the absolute deviations: 207.4

The absolute deviation for the second solution:
- » The solution: 26, 26, 66, 74, 85, 88, 90, 93, 95, 96
- » The deviations: 47.9, 47.9, 7.9, 0.1, 11.1, 14.1, 16.1, 19.1, 21.1, 22.1
- » Sum of the absolute deviations: 207.4

Students may suspect that the sum of the absolute deviations will be the same for all solutions. The following solution shows that this is not true! (Why did it change this time?)
- » The solution: 26, 26, 62, 74, 85, 88, 93, 94, 95, 96
- » The deviations: 47.9, 47.9, 11.9, 0.1, 11.1, 14.1, 19.1, 20.1, 21.1, 22.1
- » Sum of the absolute deviations: 215.4

What the sum of the absolute deviation tells you about a data set: The sum of the absolute deviations provides information about the spread (*variability*) of a data set, because it tells you how far the numbers are from the mean. Data sets that are more "spread out" have greater absolute deviation sums. In practice, people use the *mean absolute deviation* (MAD) rather than the sum. To calculate the MAD, simply divide the sum of the absolute deviations by the number of data items. For example, the MAD of the previous data set is 21.54 (215.4 ÷ 10). Thus, the MAD is the "mean of the absolute deviations from the mean"!

I wonder why the MADs are equal for so many of the solutions to this problem?

Problem #4

Mean:	73.9	Median:	86.5
Mode:	26	Range:	70

You can find deviations from numbers other than the mean.

Directions

- Choose a solution. Find the sum of the absolute deviations from the median.
- For the same solution, find sums of absolute deviations from numbers other than the median.
- Describe your observations.

Diving Deeper

Explore sums of the squares of deviations of numbers from the mean. Then explore sums of squares of deviations from numbers near the mean. Describe your observations.

CONVERSATION STARTERS FOR #4

What do you notice? What do you wonder?

In these Conversation Starters, all of the data sets are assumed to contain 10 numbers, and M represents the median of the data set.

I notice that half of the deviations from M are greater than or equal to 0 and half are less than or equal to or 0.

I notice that when I find the absolute deviations from a number other than M, each individual deviation increases or decreases by the same amount.

I notice that the sum of the absolute deviations from any number between the fifth and sixth numbers is equal to the sum of the deviations from M.

I notice that when I find the absolute deviations from a number just less than the fifth number or just greater than the sixth number, six of them increase and only four of them decrease.

I notice that as I continue to move farther from the median, the sums of the absolute deviations increase.

SOLUTIONS FOR #4

The answers below are based on the solution 26, 26, 66, 74, 86, 87, 90, 93, 95, 96. Students' answers will vary based upon the solution that they choose.

Sum of the absolute deviations from the median (86.5):
- » Absolute deviations from the median: 60.5, 60.5, 20.5, 12.5, 0.5, 0.5, 3.5, 6.5, 8.5, 9.5
- » Sum of the deviations: 183

Sum of the absolute deviations from 86:
- » Absolute deviations from 86: 60, 60, 20, 12, 0, 1, 4, 7, 9, 10
- » Sum of the absolute deviations: 183

Sum of the absolute deviations from 85:
- » Absolute deviations from 85: 59, 59, 19, 11, 1, 2, 5, 8, 10, 11
- » Sum of the absolute deviations: 185

Observations: Students may observe that the sums of the absolute deviations tend to increase as they get farther from the median. In fact, the sum of the absolute deviations from the median is always less than or equal to the sum of the deviations from any other number (including the mean)!

STAGE 3

The goal in Problem #5 is to rewrite Problem #1 so that it has the fewest possible solutions. When I first created this task and worked through it myself, I got the number of solutions down to 26, which I thought was fairly good. I let it go at that, but I was curious to see if my students could do better.

A year or two later, one of my students approached me saying that she had found a way to reduce the number of solutions to 2. I was excited but skeptical; no one else had discovered a way to do this. But when she described her ideas, I realized that my thinking about the problem had been limited. I had not seriously considered the possibility of decreasing the median, thinking that it would create too many combinations for the larger numbers. But she realized that having a *much* lower median would decrease the number of solutions by making it harder to reach a sum of 739.

The following year, another student approached me claiming to have reduced the number of solutions to one! He had realized that by increasing the median to 94, it became possible for a data set to include three 26s, which made it possible to create a problem with a unique solution!

All three of these possibilities are presented in the Solutions. Perhaps you or your students will discover flaws in our thinking or make new discoveries of your own. One of the exciting things about working with students is that they have a wonderful sense of the possible. Open-ended problems like this allow them to experience math in brand new ways.

What You Will Need

» Calculators

What Students Should Know

» Calculate the mean, median, range, and mode of a data set.

What Students Will Learn

» Analyze relationships between mean, median, range, and mode.
» Create and solve challenging problems about statistical measures.

Problem #5

Mean:	73.9	Median:	86.5
Mode:	26	Range:	70

As written, this problem has *many* solutions.

Directions

- Change the median so that the problem has the fewest solutions possible (but more than 0). The mean, mode, and range must remain the same.
- Find the number of solutions, and explain your thinking.

CONVERSATION STARTERS FOR #5

What do you notice? What do you wonder?

I wonder how many solutions the problem has as written?

I notice that when I increase the median, it reduces the number of possible values for the larger numbers in the data set.

I notice that when I decrease the median, there are fewer ways to achieve a sum of 739.

I wonder if I could increase the median by enough that it would be possible for the data set to contain more than two 26s!

SOLUTIONS FOR #5

A good answer: Set the median at 91. (26 solutions)

To limit the number of solutions, make the sixth number close to 96 or make the fifth number close to 26. Because the median must be great enough to create a sum of 739, begin by making the sixth number 92.

26, 26, _____ , _____ , _____ , 92, _____ , _____ , _____ , 96

This forces the three blanks on the right to be 93, 94, and 95. To ensure that the sixth number *must* be 92, choose a median of 91*. This means that the fifth number will be 90. Now you have:

26, 26, _____ , _____ , 90, 92, 93, 94, 95, 96

The sum of these eight numbers is 612. Because 739 – 612 = 127, the last two blanks must be filled with numbers between 26 and 90 that have a sum of 127.

38 + 89   39 + 88   40 + 87   41 + 86   42 + 85

The pattern continues until you get to 63 + 64. This is a total of 26 solutions.

If you chose a smaller median, there would be more options for choosing the middle two numbers (and more ways to fill in the blanks on the right), which would increase the number of solutions.

Note: Some students may choose a median of 91.5 (27 solutions). Try it!

A better answer: Set the median at 78.5. (2 solutions)

By making the median small enough, you limit the possible solutions by forcing the lower numbers to be so small that the mean barely reaches 739, even with the higher numbers at their greatest possible values.

To get a median of 78.5, begin with this:

26, 26, _____ , _____ , 78, 79, _____ , _____ , _____ , 96

and fill in with the highest numbers possible:

26, 26, 76, 77, 78, 79, 93, 94, 95, 96

This has a sum of 740, or a mean of 74.0. Bring it down to 73.9 by decreasing the 76 or the 93 by one. Decreasing any other numbers that filled a blank would force other numbers to decrease, which would make the mean too low.

There are only two solutions!

26, 26, 75, 77, 78, 79, 93, 94, 95, 96 and 26, 26, 76, 77, 78, 79, 92, 94, 95, 96

Making the two middle numbers farther apart (77, 80, etc.) does not work, because filling in the highest possible numbers gives a mean that is too small. The farther apart the two numbers are, the smaller the highest possible mean becomes.

What about medians of 78 and 79? The highest possible mean using a median of 78 is 73.7. A median of 79 allows more than 2 solutions, because the highest possible mean is 741 (compared to 740 for 78.5), leaving more ways to adjust the remaining numbers since you need to lower the total by 2 instead of 1.

The best answer: Set the median at 94. (1 solution!)

By making the median large enough, it is possible to raise the values of other numbers in the list enough that you can have three 26s, which enables other numbers to be repeated without affecting the mode!

26, 26, 26, 91, 94, 94, 95, 95, 96, 96

Suppose you decreased 91. You would need to compensate by increasing one of the other numbers. This would not be successful, because increasing one of the 26s would make it impossible to have two each of the 94s, 95s, and 96s, which you need in order to maintain the mean and the median. Trying to increase one of the larger numbers would make 26 no longer the only mode. Increasing 91 causes similar difficulties. Try it!

The only other possibility is to try two other middle numbers while keeping the median at 94. Using 93 and 95 does not work, because this creates two modes—26 and 95. 92 and 96 fail to work for the same reason.

ALGEBRA CONNECTIONS

In Problem #3, students discover that the sums of the deviations from the mean appear to be always equal to 0. Some may be able to give a verbal explanation for why this happens. Those who know more algebra may be able to express their ideas with variables. Suppose you have a set of five numbers, *a, b, c, d,* and *e,* having a mean of *M*. The sum of the deviations is:

$$(M-a)+(M-b)+(M-c)+(M-d)+(M-e)=$$
$$5M-(a+b+c+d+e)=$$
$$5M-5M=0$$

The last line follows from the fact that $M = \dfrac{a+b+c+d+e}{5}$, and, consequently, $a+b+c+d+e=5M$. Of course, all of this works only because some of the deviations are negative and some are positive.

For simplicity, I showed the proof for a set of five numbers. However, students can explain why it would continue to be true for data sets of 10 numbers or any other size. It will eventually be easier for students to write a general proof symbolically when they have learned about the "sigma" or "summation" symbol, Σ.

In Problem #4, students explore further by examining how absolute deviations from the median compare to deviations from other numbers. Persistent students will notice that the minimum value for the sum of the absolute deviations occurs for deviations from the median. However, if you *square* the deviations first, the minimum occurs at the mean. (They may explore this idea in the Diving Deeper question.) The squares of the deviations are important, because they appear in calculating the most frequently used measure of variability, the *standard deviation*.

Students who are familiar with the concept of *absolute value* may explore a little further. First they may try to understand why the absolute value of a number is equal to the square root of its square. In symbols:

$$|a| = \sqrt{a^2}$$

The *mean absolute deviation* calculation therefore involves squares and square roots. For example, with four numbers it would look like this:

$$\frac{\sqrt{(M-a)^2}+\sqrt{(M-b)^2}+\sqrt{(M-c)^2}+\sqrt{(M-d)^2}}{4}$$

The *standard deviation** formula involves similar calculations, but in a different order:

$$\sqrt{\frac{(M-a)^2+(M-b)^2+(M-c)^2+(M-d)^2}{4}}$$

Curious students may investigate which seems to give the greater result—taking the square root(s) before or after adding. Why does this happen? Do the two answers ever come out the same?

Note: When you are finding the standard deviation of a *sample* from your data, you subtract 1 from the total number of data points before dividing. For instance, in the example above you would divide by 3 instead of 4.

Exploration 2

A Day at the Races

What makes statistics different than other branches of mathematics? Think about this question for a moment before you read on. In mathematics, you apply logic to analyze patterns—patterns of number, shape, change, or otherwise. To the extent that you begin with correct information and follow sound logical procedures, you can reach unambiguous conclusions. In statistics, you cannot do this in quite the same way because of the *variability* in the original information. For example, if you want to know about the heart rates of sixth graders, you must work with an entire set of numbers to describe this one variable, because sixth graders do not all have the same heart rates. The best you can do is to make statements about typical heart rates, and there will always be some uncertainty in your conclusions.

In short, *statistics involves variability*. Ironically, when students begin studying statistics in school, the focus is often on calculating the mean and median—single values that summarize a whole collection of numbers. Of course, these values can be helpful, but important information is lost in the process of calculating them. The information that you lose has to do with the variability, which is precisely what statistics is all about!

The process of statistical investigation has a number of parts: (1) framing a question, designing an experiment to answer it, and collecting the data; (2) visualizing and analyzing the data; and (3) interpreting the data. The three stages in this activity match the three parts of this process. Unlike most activities in the Advanced Common Core Math Explorations series, in which the three stages gradually scaffold the level of depth and challenge, your students will need to complete this entire exploration if they are going to experience the whole process.

In fact, if you are going to deemphasize any portion of the activity, it will probably be Stage 1. Although it is important for students to gain experience with designing experiments and collecting data, it is not necessarily practical to do this for every statistics activity. Thus, in Stage 1, you have the option of asking students to collect their own data or of using the data provided on the Car Races Data Handout on page 45.

Of course, if your students generate their own data, the answers in the Solutions section will not match their work. This will require more time and effort on your part, but the Solutions will still help you by showing the kind of thinking that students should be doing.

If your students design their own questions, the teacher's support features in this exploration will be a general template rather than a detailed plan for you to fol-

DOI: 10.4324/9781003232780-5

low. The statements of Problems #1 through #5 will be virtually unchanged except that the parts referring to the car races will be replaced by your students' questions! Just be sure that their questions relate to the variability of a single numeric quantity. Otherwise, they will discover that the tasks in the problems (creating histograms, for example) will not make sense.

STAGE 1

You may use this exploration in different ways. Students may set up and run their own car races and collect their own data, or they may use the sample data set on the Car Races Data Handout. The numbers on this handout are taken from actual data that students in my classes have generated. I have modified a few values in order to bring out certain concepts more clearly, but the big picture of the results is essentially unchanged.

If your students do carry out their own experiments, you can prepare by reading through the Conversation Starters and the Solutions for Problem #1 in advance. This will help you guide them in setting up the experiment. Encourage them to choose toy cars that are different in interesting ways, especially in size or shape. Then, ask them to make predictions about which will be fastest. This heightens their engagement and gives them something to root for! Of course, they must be careful not to let their preferences affect the way that they conduct their measurements!

What You Will Need

- » A copy of the Car Races Data Handout for each student (if you will be using the data in this book)
- » Books and boards to make ramps (if you will be doing the experiment)
- » Toy cars (if you will be doing the experiment)
- » Stopwatches (if you will be doing the experiment)

What Students Will Learn

- » Design a statistical investigation (or discuss how to design one).
- » Collect and record the data (or discuss how the data on the handout may have been collected).

Problem #1

Leah, Sasha, and Jorgen each have a toy racecar, and they would like to choose the fastest one to enter in a school-wide competition. They decide to test the cars in advance by rolling them down a ramp.

Directions

Option 1:
- Look at the data on the Car Races Data Handout.
- Explain how the experiment might have been done, including the procedures for collecting and recording the data.

Option 2:
- Design and carry out an experiment with actual toy cars; collect and record your own data.
- Describe the design of your experiment in detail, including your procedures for collecting and recording the data.

For both options:
- No experiment is perfect. Discuss things that could negatively affect the accuracy or precision of the data.

CAR RACES DATA HANDOUT

S: Student number of the starter **T:** Student number of the timer

T	S	Jorgen's car (sec)	Sasha's car (sec)	Leah's car (sec)
1	2	2.01	3.23	1.91
2	3	2.06	3.05	1.95
3	4	1.85	2.88	2.38
4	5	2.25	3.70	2.20
5	6	1.96	2.50	2.01
6	7	2.23	2.98	2.00
7	8	2.16	2.73	2.18
8	9	1.83	2.53	2.15
9	10	1.86	3.20	2.35
10	11	2.11	2.85	2.28
11	12	2.20	3.35	2.18
12	13	2.05	2.40	2.23
13	14	2.46	2.26	1.96
14	15	1.80	2.83	1.76
15	16	2.06	2.48	2.05
16	17	1.55	2.40	2.30
17	18	1.91	2.46	2.13
18	19	2.13	2.58	2.38
19	20	1.96	3.06	2.16
20	21	2.03	2.55	2.16
21	22	2.70	2.35	2.25
22	23	1.85	2.96	2.03
23	24	2.03	4.05	2.00
24	25	2.01	2.07	2.20
25	26	2.08	1.38	2.14
26	27	1.80	1.66	1.85
27	28	1.81	2.52	1.50
28	29	1.93	1.91	1.71
29	30	1.85	2.11	1.43
30	1	2.18	1.81	2.10

CONVERSATIONS STARTERS FOR #1

What do you notice? What do you wonder?

I wonder how steep we should make the ramp?

I wonder how many times we should do the experiment?

I wonder if it is important to make the same number of measurements for each car?

I notice that it is important to release the cars from exactly the same spot each time.

I wonder how we can tell exactly when the car reaches the bottom of the ramp?

I wonder how precisely we should measure the times?

I wonder what details we should keep track of on paper when we record the times?

I notice that it will help to let different people release the car and measure the time each time.

I wonder what kinds of things might cause variability in the measured times?

If gravity were the only factor involved, all of the times should be nearly the same. However, the cars may have different aerodynamic profiles, causing differences in air resistance. There may also be friction in the wheels or other moving parts. You may also find that the cars do not always travel in a straight line. (My students usually consider the ability of the car to follow a straight path to be part of its "raceworthiness." Therefore, we record all time values, regardless of the path of the car.) Also, there may be small differences in the ways different people release the car and measure the times.

SOLUTIONS FOR #1

Designing the experiment: Students can make ramps by using books to prop up one side of a board. They release each car from the same position at the top of the ramp, and they use a stopwatch to measure the time it takes to reach the bottom.

A sample set of results is shown on the Car Races Data Handout. Students should take steps to get the best possible data. Some possibilities include:

- » Repeat the experiment many times for each car.
- » Let one person release the car while another person uses the stopwatch.
- » Rotate the two tasks between different students (in case one person tends to consistently measure times that are too long or too short).
- » Keep track of who makes each measurement.
- » Make a mark near the top of the ramp to ensure that the front of each car is in the same place every time.
- » Ensure that the ramp is set to the same steepness if you have to continue the experiment the next day.
- » Record times to the hundredth of a second (the accuracy shown on the stopwatch).
- » Make the ramp fairly flat in order to make the total times longer. (This makes it easier to distinguish differences in speeds.)
- » Signal the timekeeper in a consistent way each time you release a car.
- » Record the data in a neat and organized way so that no information is lost or misread.
- » Have multiple people record the data so that they can check each other's results.

The data: See the Car Races Data Handout on page 45. The variability of the times within and between cars may seem surprising! This can make an interesting topic for discussion. (See the Conversation Starters for some ideas.)

Possible imperfections in the data collection procedures:

- » There may be a small difference in the steepness of the ramp from one day to the next.
- » There may be a delay between the times when the car starts or finishes and when the recorder presses the stopwatch button.
- » The results may be affected by subtle differences in the way students release the cars at the top of the ramp. Some students may gradually become more skillful at this!

Students who are doing the experiment should consider their experimental design carefully and watch for other potential sources of inaccuracy.

STAGE 2

The focus in Stage 2 is on visualizing and analyzing the data from Stage 1. Before beginning Problem #2, check that your students are familiar with some important vocabulary. Numbers that describe a whole set of data are often called *summary statistics*. These include the mean, median, mode, range, interquartile range (IQR), mean absolute deviation (MAD), and standard deviation. The mean and the median (and perhaps the mode) are measures of *central tendency*, because they describe a typical value in the data set. The other summary statistics are measures of *variability*. They describe how "spread out" the data are.

In Problem #4, resist the temptation to teach students specific methods for estimating the mean and median from a histogram. Instead, treat it as a problem: Let students develop their own methods. As they try this and discuss their different approaches, they will develop a deeper understanding of histograms.

The Appendix at the end of this book contains examples of creating histograms and box plots, as well as calculating the IQR and MAD. Standard deviation is not used in this book, because its significance is not all that clear until students have learned more about statistical theory. However, I touch on standard deviation in the Algebra Connections at the end of Exploration 1, and the MAD makes a good bridge to understanding it. Some students may be interested in studying about it on their own!

What You Will Need

>> Graph paper and calculators

What Students Should Know

>> Plot decimal values on a number line.
>> Calculate mean, median, and range of data.
>> Create line plots (also known as dot plots) and histograms.
>> Recognize symmetric, left-skewed, and right-skewed distributions.

What Students Will Learn

>> Choose appropriate representations of data. Describe and compare them.
>> Create box plots (also known as box-and-whisker plots).
>> Calculate interquartile range (IQR) and mean absolute deviation (MAD).
>> Determine if data points are outliers.
>> Analyze relationships between graphs and summary statistics.

Problem #2

One way to understand a set of data is to summarize information about its central tendency and variability.

Directions

- Calculate some summary statistics that measure the central tendency of your data.
- Calculate some summary statistics that measure the variability in your data.

CONVERSATION STARTERS FOR #2

What do you notice? What do you wonder?

I wonder what are the best measures of central tendency for this experiment?

Try calculating both the mean and the median. Later, it may be easier to tell how they relate or which is more helpful.

I notice that the mode is probably not very useful for this experiment.

There are many different numbers in the data, and few of them are repeated.

I notice that there appears to be more variability in the data than I expected.

I wonder what are the best measures of variability for this experiment?

Consider finding the interquartile range and the mean absolute deviation in addition to the range. Think carefully about what each of these numbers tells you.

I wonder why you take the absolute values of the differences when you calculate the mean absolute deviation?

Try doing the calculation without taking absolute values. Something interesting happens that will help you understand what the absolute values mean and why they are important. (Also, see Exploration 1: Playing With Data, Problem #3.)

I notice that the mean and median are usually close to each other.

I wonder what causes the mean to be sometimes greater and sometimes less than the median?

SOLUTIONS FOR #2

Measures of central tendency:

Jorgen's Car	Sasha's Car	Leah's Car
Mean: 2.024 sec	Mean: 2.628 sec	Mean: 2.064 sec
Median: 2.02 sec	Median: 2.54 sec	Median: 2.135 sec

Measures of variability:

Jorgen's Car	Sasha's Car	Leah's Car
Range: 1.15 sec	Range: 2.67 sec	Range: 0.95 sec
IQR: 0.28 sec	IQR: 0.63 sec	IQR: 0.24 sec
MAD: 0.158 sec	MAD: 0.447 sec	MAD: 0.178 sec

Sample calculation procedures (for Jorgen's data):

» Mean: The sum of Jorgen's times is 60.71 seconds:

$$\frac{60.71}{30} \approx 2.024$$

» Median: The two middle values in Jorgen's ordered list are 2.01 and 2.03 (the 15th and 16th numbers in the list). Their mean (halfway between them) is 2.02.

» Range: maximum − minimum = 2.70 − 1.55 = 1.15

» Interquartile Range (IQR): upper quartile − lower quartile = 2.13 − 1.85 = 0.28

» Mean Absolute Deviation (MAD): The absolute value of the difference of each time from the mean:

$$|2.024 - 2.01| = 0.014$$

$$|2.024 - 2.06| = 0.036$$

$$|2.024 - 1.85| = 0.174, \text{etc.}$$

» The mean absolute deviation is the mean of these 30 distances.

» The sum of the distances is 4.75. The mean is:

$$\frac{4.75}{30} \approx 0.158$$

Problem #3

One way to understand a set of data is to visualize it.

Directions

- Decide on at least two methods for graphing your data. Explain your reasoning.
- Create the graphs.
- Discuss the general appearance of your graphs, and identify their important features.
- Talk about how you can predict the appearance of one type of graph from another. Explain your thinking.
- Determine if any of the "extreme" values are outliers. Explain.

Diving Deeper

Experiment with different bin sizes* for your histograms. Which sizes do you think produce the most useful graphs? Why?

*Note:The *bin size* is the width of each bar in the histogram (measured on the number line).

CONVERSATION STARTERS FOR #3

What do you notice? What do you wonder?

I notice that the data are not related to finding parts of a whole.

This suggests that circle graphs are not helpful.

I notice the data are not about trends (changes over time).

The data is collected over time, but the order in which the measurements occur is not relevant to the question of which car is the fastest. In fact, you would expect the variation from one measurement to the next to be random. All of this suggests that line graphs are not helpful.

I notice that only one variable is being measured.

The cars' times are the only variable. This means that a scatterplot will not be helpful.

I notice that I am taking multiple measurements of a single quantity.

Therefore, the data will have a distribution. Dot plots, histograms, and box plots are good visualization tools for distributions.

I wonder if it makes sense to make a dot plot when few values are repeated?

A dot plot may still help you visualize the distribution.

I wonder how the shape of the distribution relates to the values of the mean and median?

Which one is affected more strongly by extreme values, the mean or the median?

I notice that putting one graph above the other makes it easier to compare box plots to histograms (or dot plots).

SOLUTIONS FOR #3

Choosing types of graphs: When you have multiple values for a single variable, you want to understand how the values are distributed. Line plots, histograms, and box plots are generally the best tools for this.

Circle graphs, bar graphs, line graphs, and scatterplots are not helpful in this case: Line graphs show trends over time; circle graphs display parts of a whole; bar graphs are helpful mainly when you have a categorical (non-numeric) variable; and scatterplots are used to compare two variables. None of these circumstances apply here.

Line plots: Some students may feel that dot plots are not practical, because few numbers are repeated. However, if they experiment with plotting the points on number lines, they may discover that dot plots actually give them a useful picture of the distribution of the data.

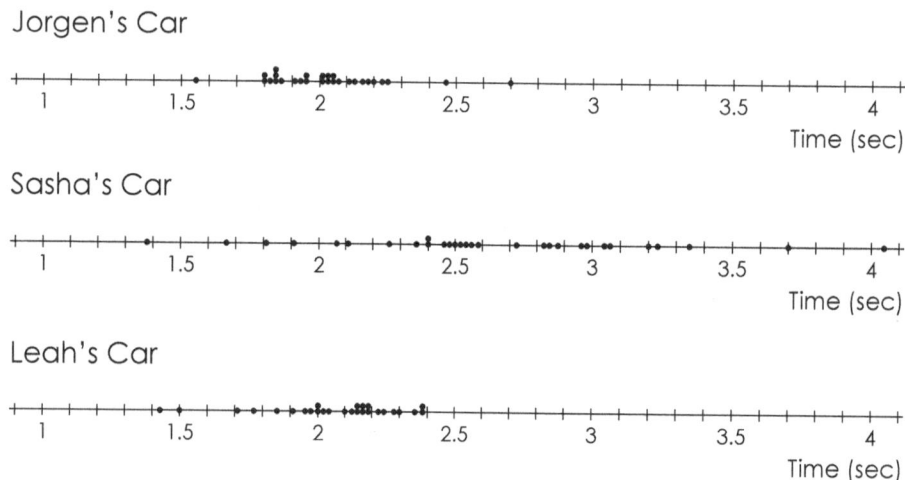

Description of the line plots: The most striking feature of the graphs is that Sasha's times show far more variability than the others. Both her minimum and maximum values are more extreme than any of the others. Her data do not cluster nearly as much, and there are more gaps between the times.

The times for Leah's and Jorgen's cars tend to cluster at values just greater than 2 seconds. Leah's times appear to cluster at slightly higher values. She also has a larger number of lower values on the left of the graph.

The data for both Leah and Jorgen contain a few surprisingly low values, and Jorgen has one or two longer times that do not seem to fit well with the rest of his data.

Histograms:

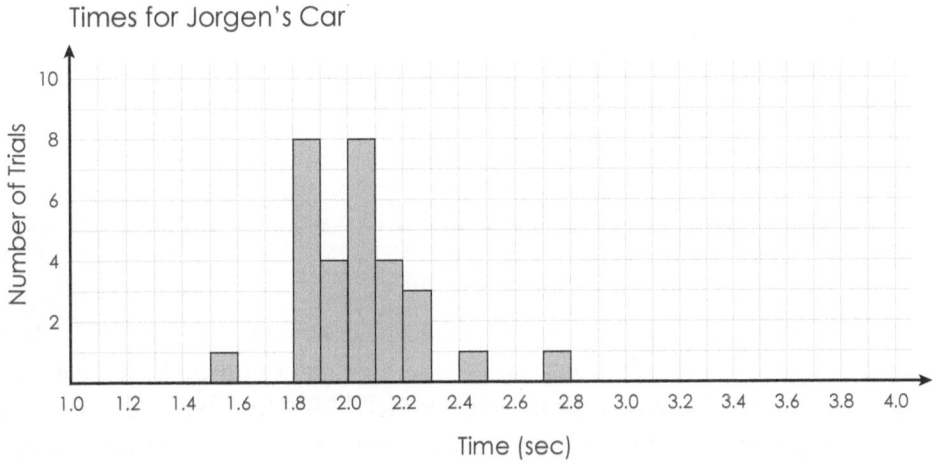

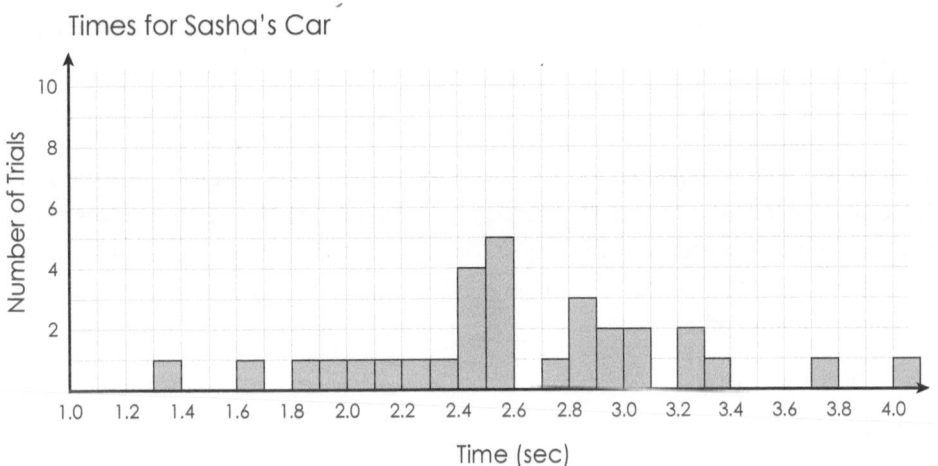

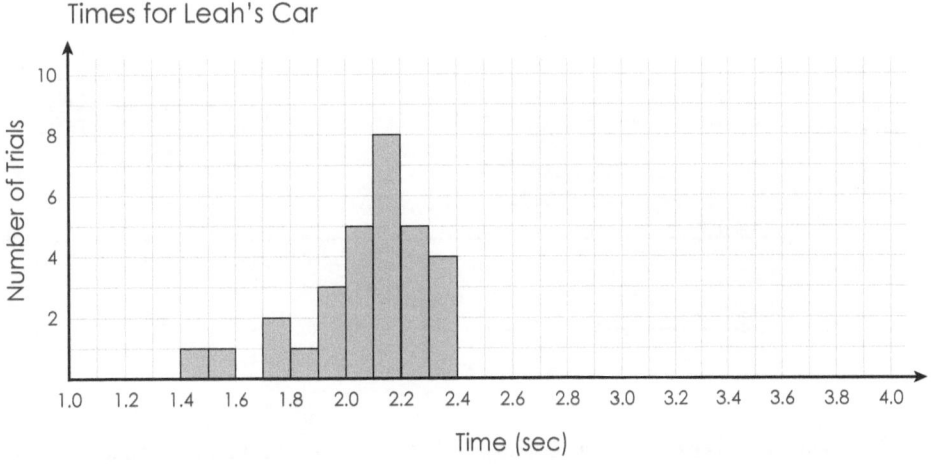

Description of the histograms: The histograms lead to the same general observations as the line plots. They do not show each data point, but, as a consequence, they are somewhat easier to read. Jorgen's and Sasha's graphs appear to be slightly skewed to the right.

Box plots:

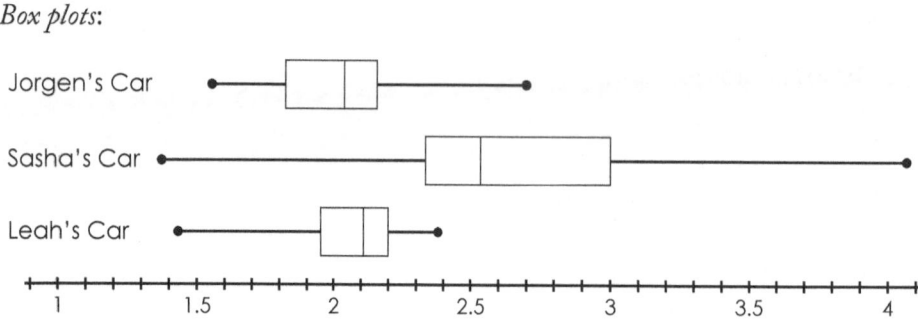

Description of the box plots: The box plots clearly show that not only are Sasha's times much more variable, but her typical values are significantly higher than either Jorgen's or Leah's. The box in Leah's plot is slightly smaller and to the right of Jorgen's, suggesting that her times are slightly longer but somewhat more consistent than Jorgen's. The longer whisker to the left of her box shows that her shorter time values are often significantly less than her typical times, while the smaller whisker on the right shows that her longer times are relatively close to her typical times.

Comparing types of graphs: Each of the four regions in a box plot contains the same fraction (one fourth) of the data. Therefore, the smaller regions are the places where the data tend to cluster, and the larger regions are places where the data are more spread out. You can see this clearly on Leah's graphs if you show the box plot and the histogram together.

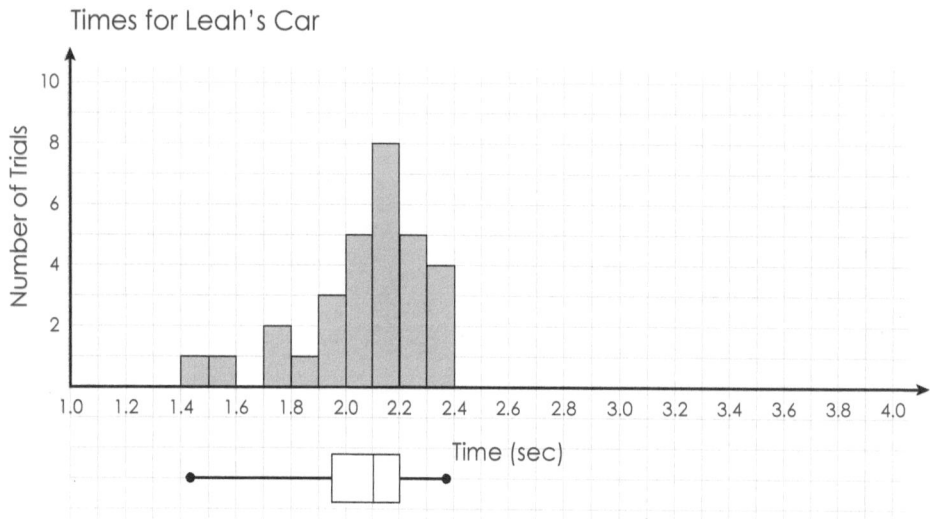

The long whisker on the left corresponds to low bars and a gap—places where the data points are spread out. The right side of the box matches with the tallest bar on the graph—a place where the data are packed closely together. You can see a similar effect if you compare the box plot to the dot plot. For example, the right side of the box will be directly below a region where the dots are clustered very tightly.

Checking for outliers:

Name	Data Point	Outlier?
Jorgen	2.70	Yes
Jorgen	1.45	No
Sasha	4.05	Yes
Sasha	1.38	Yes
Leah	1.43	Yes
Leah	1.50	Yes

Sample calculations (for Jorgen):
- » IQR = upper quartile – lower quartile = 2.13 – 1.85 = 0.28
- » IQR · 1.5 = 0.28 · 1.5 = 0.42
- » Upper quartile + 0.42 = 2.13 + 0.42 = 2.55; 2.70 is an outlier.
- » Lower quartile – 0.42 = 1.85 – 0.42 = 1.43; 1.55 is not an outlier.

Students may notice that they can estimate some of the outliers visually from the box plots. Simply decide whether the number they are testing is more than one-and-a-half "box lengths" from one end of the box!

Problem #4

You can tell some interesting things about measures of center and variability by looking carefully at histograms.

Directions

- Find a way to estimate the mean and the median of each data set directly from its histogram.
- Discuss how the shapes of the distributions relate to the summary values. For example, is it possible to predict whether the mean or median will be greater based on the shape of the graph?

Diving Deeper

Show how to use a histogram to estimate a five-number summary and draw an approximate box plot for one of the cars.

CONVERSATION STARTERS FOR #4

What do you notice? What do you wonder?

I notice that each square on the histogram's grid represents one measurement.

I notice that the median is related to the area under the histogram.

I wonder how I can estimate a value in the middle of a bar on the histogram?
    Imagine filling in each square on the grid with a realistic measurement, starting with lower numbers on the bottom of each bar. Then think about what fraction of the way up the bar you are.

I wonder how to estimate values in a bar?
    What is the most "typical" value in each bar?

I notice that I was able to estimate the mean and median pretty well even without looking at the actual data.

I wonder why my estimates for both the median and the mean are usually a little bit high?
    What is it about the actual values within each bar that causes this? What would cause the estimates to be consistently low?

SOLUTIONS FOR #4

Estimating medians from a histogram: You can estimate the median for each car to within about 1 or 2 hundredths of its actual value.

Name	Estimated Median (sec)	Actual Median (sec)
Jorgen	2.025	2.02
Sasha	2.56	2.54
Leah	2.125	2.135

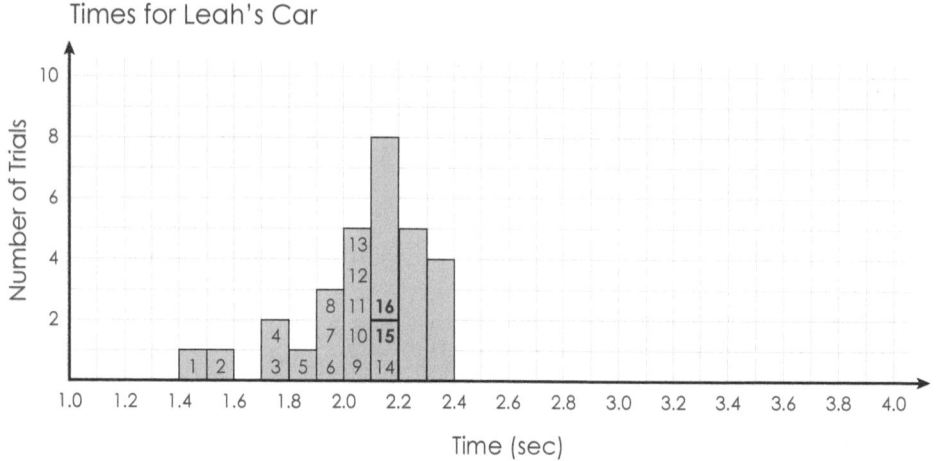

To estimate a median, simply count the squares in the bars. Look for the 15th and 16th squares. Because the location between the 15th and 16th squares sits at one fourth of the height of the bar, you might estimate a median about one fourth of the way between 2.1 and 2.2, or about 2.125. This is a little less than Leah's actual value of 2.135, because the actual data values lie a bit closer to the upper end of the interval.

Notice that what you are doing is essentially splitting the area underneath the histogram into two equal parts! In the future, as students begin to study *continuous* distributions, the idea of focusing on area will become very useful.

Estimating means from the histogram: You can estimate the mean for each car to within about 1 or 2 hundredths of the actual value.

Name	Estimated Mean (sec)	Actual Mean (sec)
Jorgen	2.037	2.024
Sasha	2.637	2.628
Leah	2.077	2.064

A sample calculation (for Leah): Refer again to the histogram above. Begin by approximating each bar by an intermediate value. For example, you can approximate the bar between 2.0 and 2.1 as 2.05. For Leah, this bar is 5 units tall, meaning that there are five values between 2.0 and 2.1. Their sum is approximately $2.05 \cdot 5 = 10.25$. Do this for all of the bars, and estimate the sum by adding the results. To estimate the mean, divide by 30 (the sum of the heights of the bars):

$$1.45 + 1.55 + (1.75 \cdot 2) + 1.85 + (1.95 \cdot 3) + (2.05 \cdot 5) +$$
$$(2.15 \cdot 8) + (2.25 \cdot 5) + (2.35 \cdot 4) = 62.3$$

$$\frac{62.3}{30} \approx 2.077$$

This estimate is a little bit greater than the actual mean, because the values within each bar tend slightly toward the lower end of the interval.

Relating the shapes of the distributions to the means and medians: Jorgen's and Sasha's graphs appear to be slightly skewed to the right. This is reflected in the summary statistics; the means for their data are a little bit greater than the medians. This makes sense, because the higher values affect the mean more strongly than the median. Leah's graph, on the other hand, is skewed somewhat to the left, causing her mean to be a little less than her median.

STAGE 3

By now, the students have spent a lot of time visualizing, analyzing, and perhaps collecting data. However, the most important part still remains! The purpose of the earlier work was to answer the original question: that is, to draw conclusions and make decisions.

What You Will Need

» Copies of your data, calculations, and graphs from Stages 1 and 2

What Students Should Know

» Understand concepts and skills from Stages 1 and 2.

What Students Will Learn

» Use graphs and summary statistics to interpret data and draw conclusions.
» Recognize and describe ways in which procedures for collecting data may affect conclusions.

Problem #5

In the end, the purpose of carrying out experiments and collecting data is to answer questions in the real world!

Directions

- *Interpret* your data*. In other words, based on your analysis of the data, decide which car should be entered in the competition. Explain your reasoning. Make use of your graphs and summary statistics.
- Think about how your data were collected. Does this have any bearing on your interpretation? Explain.

Note: *Interpreting* data is about telling what the data mean—using your graphs and numbers to answer the original question.

CONVERSATION STARTERS FOR #5

What do you notice? What do you wonder?

I notice that answering the original question about choosing a car involves a lot more thought than just describing the graphs and analyzing the summary statistics.

I notice that it is not enough just to focus on the mean and the median.

I wonder how important the variability is in deciding which car to choose?

I notice that the detailed shapes of the distributions could make a difference in my decision.

I notice that knowing how the competition is going to be judged could affect my decision.

I wonder if there was anything in the way the data were collected that might make all of the measured times longer (or shorter) than they really were?

I wonder if there was anything in the way the data were collected that might affect the measured times differently for one car than for another?

I wonder if the differences in the measurements for Leah's and Jorgen's cars reflect actual differences between the cars or just random uncertainties in the measurements?

This is an important question. It will be addressed in Exploration 4 when students explore sampling distributions.

SOLUTIONS FOR #5

Making a decision: Sasha's car is clearly the slowest overall and the least predictable. Her mean and median are much greater than Jorgen's and Leah's, as are her measures of variability. You can also see this by looking at the histograms and box plots. The bars in the histogram and the box in the box plot show the data clustering farther to the right; the dots are more spread out, and the box is much wider.

The distinction between Leah's and Jorgen's cars is not quite as clear. Jorgen's car appears to be faster. His mean and median times are both a little less than Leah's, and the box in his box plot is slightly to the left of hers. However, the measures of central tendency do not tell the whole story.

Leah's data shows a little less variability. The box in her box plot is narrower, showing that her interquartile range is less than Jorgen's. Her range is also smaller. If consistency and dependability are important, her car appears to have a slight edge.

You might argue that Leah's apparent advantage in dependability is questionable, because her mean absolute deviation is greater than Jorgen's. However, it is interesting to look more closely at the details. The values that are far from the mean in Leah's data tend to be to the left on the number line. (This is suggested by the string of dots on the left side of her dot plot, the long whisker on the left side of the box in the box plot, and the fact that her mean is less than her median.) In other words, much of the variability in her data seems to be attributable to faster times. The situation is the opposite for Jorgen. His data is skewed somewhat to the right as evidenced by the long whisker on the right and a mean that is greater than the median.

In practice, it may not matter too much which of the two cars you choose for the competition. Interestingly, I find that many students tend to prefer a more dependable car. These students may be inclined to choose Leah's car, especially because much of her variability is attributable to faster times.

In the end, the choice may depend on how the racing competition is set up and on the contestants' tolerance for risk. Do the cars at the competition race just one time? If so, perhaps Jorgen's car is the best bet. On the other hand, if they race multiple times and the fastest time is taken, Leah's car may be preferable.

In fact, in this case, some people may even choose Sasha's car! It did have a number of fast times, and it had the smallest minimum—the single fastest time—of any car! However, this number (1.38 seconds) was an outlier. It is possible that it came from a measurement error and that her car did not really go that fast.

How data collection procedures may affect the interpretation: Ideally, if data are collected and recorded with thought and care, the procedures will not have much of an impact on your interpretation. However, it is an important issue to consider. For example, you may notice clusters of shorter times near the end of the data collection process. This may suggest the possibility that students (even unconsciously) became gradually better at releasing the cars—perhaps more cleanly or in a way

65

that helped them follow a straight path. Although this may seem unlikely, it is important to be aware of potential factors in the data collection process that could influence your results.

Scientists, in particular, have to learn to watch for possible sources of *systematic error*—that is, variability that biases the data consistently in one direction. For example, an electronic instrument (or your stopwatch!) may be calibrated incorrectly so that it measures values that are always greater than the actual values. Sources of systematic error can be notoriously challenging to identify and may lead to incorrect conclusions.

Exploration 3
Simulation Station

Simulating an action is like pretending to do it. Many simulations serve a very useful purpose. For example, a flight simulator helps you learn to fly a plane without getting into a real plane. Surgical simulators are being created to help train doctors. Students may be familiar with simulation-based computer games. And virtual reality technology can simulate all sorts of situations that may or may not be practical (or desirable!) to experience in actual reality.

Mathematicians and scientists use simulations to model the real world and make predictions. For example, meteorologists create mathematical computer models to simulate weather or climate conditions and make forecasts. Simulations are also used to estimate probabilities that may be too difficult (or even impossible) to calculate directly.

When students are first learning about probability, simulations help them make the link between mathematical calculations and the real world. Students may also apply their new knowledge of simulations to test their calculations in upcoming explorations.

In Simulation Station, students have many opportunities to create their own simulations in order to estimate probabilities. They will have the advantage of being able to test their predictions, because their probabilities will *not* be impossible to calculate (although some of them may be challenging)! As they determine probabilities using both simulations and calculations, they develop a deeper understanding of the meaning of probability along with tools and strategies to hone their computational skills.

I always begin by making a set of tools available to students, including dice, coins, and playing cards. The key feature these tools have in common is that they can produce *random* outcomes—that is, outcomes that are equally likely. I generally ask students to choose which tools to use and how. They will learn at least as much from their unsuccessful attempts as they do from their successes!

DOI: 10.4324/9781003232780-6

STAGE 1

In the three Stage 1 problems, students design and carry out simulations in order to estimate probabilities involving sports, weather, and games. When practical, they compare their predictions to expected results. Before beginning, you will want to introduce or review some basic vocabulary. An *experiment* is a set of one or more actions (such as flipping coins or rolling dice) with an identifiable result. Each result is called an *outcome*. If two or more outcomes have the same probability of occurring, they are called *equally likely outcomes*. Your students may not completely understand these terms right away, but their understanding will develop as they use them in context.

You may introduce other terms as they arise. For example, *experimental probability* is an estimated probability obtained by performing an experiment. *Theoretical probability* is the calculated probability based on a mathematical model. When the outcomes are equally likely (as they are in Stage 1), the theoretical probability (P) is given by the formula:

$$P = \frac{\text{number of outcomes in the event}}{\text{number of outcomes in the sample space}}$$

The *event* is the set of outcomes you are considering. For example, if you are finding the probability of getting a head and a tail (in either order) when you flip a coin twice, then the event consists of the outcomes "HT" and "TH." (Often, an outcome in the event is called a *favorable outcome*.) The *sample space* is the set of all possible outcomes.

Note: Other terms, such as *fair game*, are defined in the Conversation Starters or the Solutions as needed. In Exploration 6, students will learn more formal terms for some of these concepts as they begin to explore them in more depth.

What You Will Need

» Coins, dice, playing cards, or other tools for performing simulations

What Students Should Know

» Understand probability as a number between 0 (impossible) and 1 (certain).
» Express fractions as percentages.

What Students Will Learn

» Design and perform simulations in order to find experimental probabilities.
» Develop strategies (such as making lists or tables) for organizing outcomes.
» Begin calculating theoretical probabilities when outcomes are equally likely.
» Understand the relationship between theoretical and experimental probability.

Problem #1

The Flyers and the Gliders are playing a baseball double-header. The two teams are equally matched.

Directions

- Predict the probability that the teams will split the doubleheader. Explain your thinking.
- Design and carry out a simulation to estimate the probability. Describe the simulation process in detail.
- Decide whether your simulation supports your prediction. Explain your reasoning.
- Repeat the directions above for a three-game series in which you find the probability that the same team will win all three games.

Diving Deeper

Although this may not be realistic from a baseball point of view, try extending your mathematical investigation to a series of more than three games. Do you see any patterns?

CONVERSATION STARTERS FOR #1

What do you notice? What do you wonder?

I notice that equally matched teams should each have a 50% probability of winning.

I wonder what it means to split a doubleheader?

It means that each team wins one game.

I wonder if it matters who wins which game when they split the doubleheader?

Yes, it does matter in the sense that FG and GF are different outcomes. (See the Solutions.) Try to allow students to reach this conclusion on their own through discussion and by looking at the results of their simulations.

I wonder what tool I should use for my simulation?

Coins are a natural choice, but you could use dice, spinners, or many other tools as long as there are two equally likely outcomes.

I wonder if it matters whether I flip two coins at the same time or the same coin twice?

I wonder how many times I should do the experiment?

Encourage students to observe the effects of doing more and more experiments. Some will understand intuitively that the more times you do it, the more confident you can be in the result.

I wonder if a team's momentum from the first game could affect the probabilities?

The simulation and the calculations are based on assuming that the probability for each team remains 50% at all times. However, it is possible that momentum or other emotional factors could change the 50% number once the games have begun.

SOLUTIONS FOR #1

Prediction for the probability: Students' predictions may vary. The correct value is 50%.

It comes from recognizing four outcomes: FF, FG, GF, and GG. For example, FG represents an outcome in which the Flyers win the first game and the Gliders win the second. Of the four equally likely outcomes, two (FG and GF) fit the scenario in which each team wins one game. Therefore, the probability is $\frac{2}{4}$ or 50%.

Students often predict $\frac{1}{3}$, because they believe that FG and GF are the same outcome. If this happens, do not tell them which prediction is correct! Instead, use the simulation to gather evidence. (Hopefully, the evidence will support 50%!)

The simulation:
»   Choose an action to simulate playing a game. (A natural choice is flipping a coin, because there are two equally likely outcomes for both the coin and the games.)
»   Flip the coin twice. (This is one *experiment*.)
»   Mark the outcome as *favorable* (for FG or GF) or *not favorable* (for FF or GG).
»   Repeat many times.
»   Divide the number of favorable outcomes by the number of experiments.

Deciding if the results support the prediction: Suppose that the students did the experiment 20 times with these results:

FG	GG	FF	GG	**GF**	FF	GG	**GF**	GG	**GF**
GG	**GF**	FF	**FG**	GG	**GF**	**GF**	FF	**FG**	FF

9 of the 20 outcomes are favorable (shown in bold). $\frac{9}{20}$ is closer to $\frac{1}{2}$ than to $\frac{1}{3}$, which suggests that the 50% prediction is more likely to be correct. Students should understand that the more times they perform the experiment, the more confident they can generally be that the result will be close to the actual (*theoretical*) probability. After students have performed the simulation individually or in small groups, consider combining the class data and comparing it to the individual results.

The three-game series: The setup for the simulation is the same as before except that each experiment now involves three coin flips, and you search for outcomes in which the same side of the coin comes up each time. Some students may be able to apply what they learned earlier in the problem to make better predictions. The possible outcomes are:

FFF   FFG   FGF   GFF   GGG   GGF   GFG   FGG

Two of the eight outcomes (FFF and GGG) are favorable. Therefore, the probability that a single team will win all three games is $\frac{2}{8}$ or 25%.

This is a great opportunity for students to begin discussing how they organize and represent their outcomes. Did they follow a pattern in order to prevent missing or double counting outcomes? Did they use tables, lists, or tree diagrams? (These are discussed in more detail in later explorations.)

Problem #2

There is a 50% probability of rain on Saturday and a 50% probability of rain on Sunday.

Directions

- Predict the probability that it will rain over the weekend.
- Design and perform a simulation to test your prediction.
- Redo your prediction and simulation for a 20% probability of rain on Saturday and an 80% probability of rain on Sunday.

Diving Deeper

Create an algebraic formula to calculate the probability of rain over the weekend if the probability of rain on Saturday is *a*, and the probability on Sunday is *b*. Suggestion: Express percentages as decimals. (See the Algebra Connections for more information.)

CONVERSATION STARTERS FOR #2

What do you notice? What do you wonder?

I notice that it does not make sense to add the two probabilities.

I notice some similarities to Problem #1.

I notice that there are four equally likely outcomes.

I wonder if I have to flip the coin (or perform my action) a second time if it comes up "rain" the first time?

I notice that I can think of "first flip" and "second flip" instead of "rain on Saturday" and "rain on Sunday."

I wonder if the weather on Saturday will affect the weather on Sunday?

This is a subtle question. We are assuming that the probabilities for the two days are *independent*—that is, that one outcome does not affect the other. However, if it does rain on Saturday, it may become more or less likely to rain on Sunday. Mathematicians often make simplifying assumptions in order to make calculations manageable. However, they must ensure that the assumptions are reasonable.

When students begin working on the task with the 20% and 80% probabilities:

I notice that one probability increased by the same amount that the other decreased.

I wonder if this means that the overall probability stays the same as before?

I notice that it is not practical to use coins as a tool for this simulation.

I wonder what tools could simulate 20% and 80% probabilities?

I wonder what it really means to say that there is a certain probability of rain one day?

SOLUTIONS FOR #2

There is a 75% probability of rain over the weekend.

Sample thinking process: There are four equally likely outcomes.

no rain Saturday no rain Sunday	rain Saturday no rain Sunday
no rain Saturday rain Sunday	rain Saturday rain Sunday

I have shown these in a table, but students may show it in many other ways. For example, they might create a list: NN, NR, RN, and RR. Because three of the four outcomes involve rain, the probability is $\frac{3}{4}$ or 75%. (Some students may begin by noticing that the probability of *no* rain is 25% and finish by subtracting 25% from 100%.)

Testing the prediction with a simulation: One approach is to use a coin.
> » Flip the coin twice. (For example, the first flip may represent Saturday and the second one Sunday.)
> » Suppose you choose to let heads stand for "rain" and tails for "no rain." If either coin comes up "heads," then count it as a favorable outcome.
> » The experimental probability equals the number of favorable outcomes divided by the number of experiments.
> » Some students may choose tools other than coins.

Prediction for the 20% and 80% scenario: The actual probability is 84%. Some students may believe that the probability will remain the same as before, because one percentage increased by 30% while the other decreased by 30%. Others may recognize that this does not make sense, because if the probabilities increased/decreased by 50%, they would be 100% and 0%, and there would certainly be rain over the weekend! A few students may be able to do a calculation to find the exact answer. (See the note below.)

Simulation for the 20% and 80% scenario: Coins will no longer work, because the probabilities are not equally likely. Students need to think of actions that have a 20% and an 80% chance of occurring. One possibility is to use 10 playing cards numbered A (Ace), 2, 3, 4, 5, 6 7, 8, 9, and 10. For example:
> » Mix the cards. Choose one without looking. If the card is A or 2, mark "rain" for Saturday. Otherwise, write "no rain."

» Mix the cards again. Chose one without looking. If the card is A, 2, 3, 4, 5, 6, 7, or 8, mark "rain" for Sunday. Otherwise, write "no rain."

» If either day is a "rain" day, mark the experiment as a favorable outcome. If not, mark it as unfavorable.

» Repeat the experiment many times.

» Divide the number of favorable outcomes by the number of experiments.

Note: Some students may calculate the theoretical probability. One method is to show all 100 pairs of cards that could be drawn. 84 of the 100 pairs correspond to rain over the weekend! Students who do this may also want to attempt the Diving Deeper question.

Problem #3

Brianna and Mahir are playing a game. They take turns rolling two standard six-sided dice. Brianna gets a point whenever the product of her two numbers is even. Mahir gets a point whenever the product of his two numbers is odd. Each person rolls eight times. The person with the most points wins.

Directions

- One of the players becomes dissatisfied with this game. Explain who and why.
- Change the rules of the game in order to make it fair. Explain your thinking.
- Play the fair game many times. Record your results and discuss whether they support your claim that the game is now fair.

Diving Deeper

- Suppose the players roll three dice instead of two. Do you need to change the rules again in order to make the game fair? Continue increasing the number of dice. Look for patterns, explain what causes them, and generalize to any number of dice.

- Who has the advantage if points are awarded based on whether or not the product is a multiple of 3? What about a multiple of 4? Is it possible to make these games fair?

CONVERSATION STARTERS FOR #3

What do you notice? What do you wonder?

I notice that it might help to begin by playing the game a few times!

I wonder if it matters who rolls the dice?
>It should not matter, because neither person can control the result of a roll.

I wonder if it matters that they roll the dice eight times?

I wonder if it matters whether each person rolls the dice one at the time or both at once?
>Again, it should not matter, because the outcome of one roll is independent of (not affected by) the outcome of another roll.

I notice that there is more than one way to change the rules.

I wonder how many points Brianna would typically be expected to have after 8 rolls?

I wonder how many points Mahir would typically be expected to have after 8 rolls?

I notice that the product of two even numbers is always even.

I wonder what happens with the product of two odd numbers? One of each?

I wonder what causes these patterns?

I wonder if I can make the game fair by changing the goal of getting an even product?

I wonder if I can make the game fair by changing the way points are awarded?

I wonder if I can make the game fair by giving a person more than one roll on each turn?

I wonder if I can make the game fair by subtracting points from a player's score in some circumstances?

SOLUTIONS FOR #3

Who becomes dissatisfied and why: Mahir eventually discovers that he loses more often than he wins. Listing all possible pairs shows that only one fourth of them (shown in bold) have an odd product!

1, 1	2, 1	**3, 1**	4, 1	**5, 1**	6, 1
1, 2	2, 2	3, 2	4, 2	5, 2	6, 2
1, 3	2, 3	**3, 3**	4, 3	**5, 3**	6, 3
1, 4	2, 4	3, 4	4, 4	5, 4	6, 4
1, 5	2, 5	**3, 5**	4, 5	**5, 5**	6, 5
1, 6	2, 6	3, 6	4, 6	5, 6	6, 6

Some students may observe that both factors must be odd if the product is to be odd. In other words, if *either* number is even, the product will be even. (Why is this?)

Changing the rules to make the game fair: In order for the game to be fair, each person must have an equal probability of winning. Of course, one way to make the game fair is for both players to get a point for the same thing (both for even products or both for odd products). Some students may suggest instead changing the word "product" to "sum" in the rules. This will work as well!

A creative possibility is to change the point system. Three out of every four products is even. Therefore, if Mahir gets 3 points for each odd product (while Brianna still gets 1 point for each even product), the game will be fair! For example, in a typical game Brianna will roll an even product 6 times, so her score will be 6. Mahir will roll an odd product 2 times, and his score will be $2 \cdot 3 = 6$—the same as Brianna's!

Playing the game: Over a short run of games, percentages of wins may vary quite a bit. However, if the revised game is fair, and if students play for quite a while, they should discover that each player wins about the same number of times when they play the revised version of the game. If the revised game is not fair, they may be able to discover this by playing!

STAGE 2

Problem #4 addresses the same concepts as the problems in Stage 1, but it is more complex. Students learn to be thoughtful when they design their simulations, because there are pitfalls that may catch them off guard if they do not think carefully. Part of the fun of the problem is that when students finish their simulations, they are often surprised at how large their answers are!

Some students may be concerned about the amount of time it will take them to do the simulation. In my experience, 200 or 300 experiments are usually sufficient, and students are often pleasantly surprised by how quickly they can do them. If the process *is* too slow, they might consider using a random number generator. These are available on many websites. I have found that the tool at http://random.org works very well.

After finishing the problem, students may enjoy combining their results to calculate an experimental probability for the entire class. This is a great opportunity to discuss the effects of increasing the total number of experiments. If you are not planning to do Problem #5 (in which students calculate the theoretical probability for this situation), you may share this percentage (approximately 23.6%) with them so that they can see how close they came as a class. If they will be working on Problem #5, they will soon discover the theoretical probability themselves!

What You Will Need

>> Coins, dice, playing cards, or other tools for performing simulations
>> A computer or other device for generating random numbers (optional)

What Students Should Know

>> Understand simulations and experimental probability at the level of Stage 1.

What Students Will Learn

>> Design and perform simulations in order to find experimental probabilities in complex situations.
>> Further develop strategies for organizing outcomes in order to count them.
>> Understand the relationship between theoretical and experimental probability.

Problem #4

Three people are chosen at random and asked their birth months.

Directions

- Design a simulation to estimate the probability that at least two of the people will have the same birth month. Describe each step in the process clearly.
- Carry out your simulation, and explain how you used it to get your result.

CONVERSATION STARTERS FOR #4

What do you notice? What do you wonder?

I wonder if each birth month is equally likely?

Not quite. For example, more babies tend to be born in July and August than in other months. However, each month is *approximately* equally likely. Mathematicians and scientists must often make simplifying assumptions in order to make a problem manageable.

I wonder how much detail I should give when I describe my simulation?

Talk about the tools you will use, how you will use them, and how you will record your results. Be very specific. One way to test your description is to give it to another person who is unfamiliar with what you are doing. Ask them to perform the simulation. If they can do it correctly, your explanation is likely to be good. If not, you may get some ideas about how to improve it.

I wonder how many times I should perform my experiment?

Try to strike a balance between being practical and being confident in your result. (Doing 200 or 300 experiments has usually been a reasonable compromise for many of my students—depending on their simulation method and how much time they have to complete the task.)

I wonder if I should express my answer as a fraction or as a percentage?

Both work well. It may be easier to compare your result to the theoretical calculation in Problem #5 if you use a percentage.

I notice that the probability is greater than I expected!

I notice some surprising "coincidences" in my list of outcomes.

This is to be expected! Many things may look like coincidences to us. For example, you may have four, five, or more favorable outcomes in a row. Or many consecutive favorable outcomes might be for the same month. Although the probability of any *particular* coincidence occurring may be small, the chance of *some* coincidence showing up in your data is fairly large—because there are so many of them!

SOLUTIONS FOR #4

Designing the simulation: Students may use spinners, playing cards, pieces of paper in a bucket, a die and a coin, a random number generator on a computer, or creative ideas of their own. An appropriate set of steps could be:

» Select an action that represents asking a person his or her birth month. It must consist of 12 equally likely outcomes.

» Perform the action three times. (This is one *experiment*.)

» If two or three of the results are the same, record the outcome of the experiment as *favorable*. Otherwise, record it as not favorable.

» Repeat this process many times.

» Divide the number of favorable outcomes by the total number of outcomes.

Carrying out the simulation: When students do a few hundred experiments, their answers often lie between 22% and 25%. The theoretical probability is $\frac{17}{72}$ or about 23.6%. Results significantly outside this range may be due to errors in the simulation process. For example:

» Students who draw slips of paper from a bucket must ensure that each slip is the same size and shape. They must also mix the slips thoroughly between each draw.

» Some students create 36 slips (or use 36 playing cards) with three slips for each month, believing that this makes it possible to draw three slips without replacing them between each draw. Help them understand that they must still replace the slips! Otherwise, the probabilities are no longer all $\frac{1}{12}$ after the first draw.

» Some students try rolling a six-sided die twice, because $6 + 6 = 12$. They do not realize at first that the number of outcomes in this case is $6 \cdot 6 = 36$. What does work is to use a die and a coin! This gives $6 \cdot 2 = 12$ outcomes. For example, heads could represent the first six months and tails the last six months of the year.

Some students may be interested in using a random number generator to choose numbers between 1 and 12. These are available on some websites (such as http://www.random.org) as well as on many graphing calculators. Some websites will generate as many numbers as you like all at once. They may even format them into columns in order to make them easier to read.

STAGE 3

Some teachers may prefer to wait until students have more experience using lists, tables, or tree diagrams to find and organize outcomes (see Explorations 5 and 6) before beginning Problem #5. I am including the problem with this exploration for three reasons. First, many students will have already explored the ideas just mentioned in their curriculum. Also, curious and persistent students may be able to solve the problem regardless! Finally, students get excited about comparing their experimental probabilities from Problem #4 to the theoretical value, and it is even more fun when they can calculate the theoretical probability themselves!

What Students Should Know

- » Have experience calculating theoretical probabilities in simpler situations.
- » Use lists or tables (or possibly tree diagrams) to organize outcomes in order to count them.

What Students Will Learn

- » Extend understanding of theoretical probability.
- » Apply knowledge of theoretical probability to solve a complex problem.
- » Understand the relationship between theoretical and experimental probability.

CONVERSATION STARTERS FOR #5

What do you notice? What do you wonder?

I notice that there are too many outcomes to list them all.

I notice that it helps to solve a simpler problem first.
     Imagine that there were only two or three months in a year! How would you count the total number of outcomes in that case?

I notice that I can start by counting the combinations for one month being shared.

I wonder if an outcome of JJF (January, January, February) is different than JFJ and FJJ?

I notice that I have to be careful how I count the outcomes in which all three people give the same answer.

I wonder how I should round my answer?
     How precise do you think the answer might be? Remember that we assumed equally likely probabilities for each month.

I wonder if there is a faster way to calculate the probability (without counting the outcomes)?

I wonder if it would be helpful to find the probability that none of the people have the same birth month?

SOLUTIONS FOR #5

The theoretical probability is $\dfrac{408}{1728}$, which is equivalent to $\dfrac{17}{72}$ or about 23.6%.

Strategy 1: Students' usual strategy is to calculate the total number of outcomes and the favorable outcomes. The total number of outcomes is $12^3 = 12 \cdot 12 \cdot 12 = 1728$. Some students may find it helpful to make tables or tree diagrams in order to discover the idea of multiplying $12 \cdot 12 \cdot 12$. (See Explorations 5 and 6 for more details.)

Counting the favorable outcomes may be more challenging. One approach is to begin by focusing on one month, say, January. In the following list, I use numbers to represent the months. Note that if the three people give the same three answers but in a different order, it is a separate outcome:

1,1,1	1,2,1	2,1,1	1,1,2	1,3,1	3,1,1	1,1,3
1,4,1	4,1,1	1,1,4	1,5,1	5,1,1	1,1,5	1,6,1	6,1,1	1,1,6
1,7,1	7,1,1	1,1,7	1,8,1	8,1,1	1,1,8	1,9,1	9,1,1	1,1,9
1,10,1	10,1,1	1,1,10	1,11,1	11,1,1	1,1,11	1,12,1	12,1,1	1,1,12

There are $1 + 11 \cdot 3 = 34$ ways in which two or three people will share a January birth month. Because this is true for each month, there are $34 \cdot 12 = 408$ favorable outcomes.

The theoretical probability is the number of favorable outcomes divided by the total number of outcomes:

$$408 \div 1728 = \frac{408}{1728} = \frac{17}{72} \approx 23.6\%$$

Strategy 2: Students who understand the *multiplication rule* for independent events (see Exploration 6) may occasionally discover the following efficient strategy!

The first person's choice does not immediately affect the calculation. The probability that the second person's birth month is the same as the first is $\dfrac{1}{12}$.

The probability that the first two people will have different birth months ($\dfrac{11}{12}$) but the third will match one of the first two ($\dfrac{2}{12}$) is:

$$\frac{11}{12} \cdot \frac{2}{12} = \frac{11}{12} \cdot \frac{1}{6} = \frac{11}{72}$$

Either situation results in a match. Therefore the probability is the sum:

$$\frac{1}{12} + \frac{11}{72} = \frac{6}{72} + \frac{11}{72} = \frac{17}{72}$$

Some students may illustrate their thinking with a simple tree diagram. M stands for "match" and N for "no match."

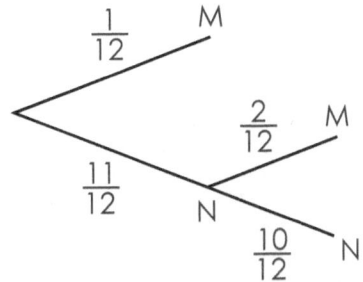

Strategy 3: This is a variation of Strategy 2. The probability that neither person will match the first person is $\frac{11}{12} \cdot \frac{10}{12} = \frac{55}{72}$. Thus, the probability that someone *will* match is:

$$1 - \frac{55}{72} = \frac{17}{72}$$

Comparing the theoretical and experimental probabilities: Assuming that they performed the experiment enough times, most students should find that their experimental calculations are within a few percent of the theoretical probability.

Exploration 4

Comparing Populations

When you ask a statistical question, it has to do with some *population* of people or things. For example, if you want to predict how the people in your state will vote on a ballot issue, then your population consists of the voters in your state. Because it is not practical to ask all of them how they plan to vote, mathematicians use a process called *sampling*. You begin by selecting a *representative sample*—a subgroup of the population that has characteristics similar to the population as a whole. Your sample must be representative in order to use its data to make useful predictions or *inferences* about the entire population.

You might think that you could create a representative sample intentionally by anticipating and balancing its characteristics. Unfortunately, you cannot know or even imagine all possible relevant characteristics. Instead, you may choose from a number of standard methods for selecting samples. As you will see, some methods are better than others at producing representative samples.

Suppose, for example, that you want to know how many text messages a typical student at your school sends each day. Your population is the set of students at your school.

» *Random sampling*: You choose the members of your sample in such a way that every student at your school has an equal chance of being selected—that is, *randomly*. (Students who worked on the Simulation Station exploration learned to use tools such as dice, coins, and spinners to generate random outcomes. These same tools are helpful in selecting random samples from a population. Random number tables or websites that generate random numbers are also helpful.)

» *Systematic sampling*: You use a predictable method for choosing your sample. For example, you might stand at the door of the school as students are entering and ask every 10th person how many text messages he or she sends each day.

» *Voluntary response sampling*: You allow students at your school to choose whether to be part of the sample. For example, you might place a questionnaire at the entrance to your school with a sign asking interested people to fill it out.

» *Convenience sampling*: You select students for your sample using a method that is convenient for you. An example of this would be asking only your friends.

DOI: 10.4324/9781003232780-7

Assuming that the size of your sample is large enough, random sampling is the process most likely to produce a representative sample. When it is not possible or practical to get a random sample, systematic sampling may be the next best choice, although it does depend on the situation and the system that you choose. Voluntary response and convenience sampling are generally not ideal choices, but in some cases they may be the only practical options.

There are two ways to use this exploration. Students may create their own questions and gather their own data. Or they may use the question and the data provided in Problem #1. It is important for them at some point to gain experience with collecting data—not necessarily in every statistics activity.

STAGE 1

The first part of this activity is about creating questions and gathering data from a sample. If your students are using the question and the data in this book, then Stage 1 may go quickly, because their task will simply be to guess how the data may have been gathered.

For students who are designing their own experiments, suggestions for creating and phrasing good questions as well as selecting samples are provided in the Conversation Starters. If your students are having trouble coming up with ideas for questions, some of the following suggestions for populations and data may be helpful.

Ideas for populations: students of certain ages at your school; adults/students; boys/girls; books by different authors, or from different genres, or for different ages; fiction/nonfiction books; different sports teams or players. Notice that your population does not necessarily have to consist of people!

Ideas for data: number of texts sent each week; number of hours slept per night; number of posters in your bedroom; number of pets in your household; perception of how much time passes within a given interval; number of siblings; number of children living on your block; number of books you read last summer; the length of your foot or arm; the number of states you have visited; number of letters per word in a book; number of sentences per paragraph, etc.

What You Will Need

- » A tool for generating random numbers (for Option 1)
- » Handout for Comparing Populations (for Option 2)

What Students Should Know

- » Understand concepts from Exploration 2: A Day at the Races.
- » Use tools to generate random outcomes (See the Simulation Station activity).
- » Understand the meaning of the word *random*.

What Students Will Learn

- » Select representative samples from a population.
- » Choose an appropriate sample size.

Problem #1

Sometimes it is not practical to gather data from the entire population that you are studying.

Directions

Option 1:

- Ask a question that compares some feature of two large populations. Identify the populations you are studying.
- Select a representative sample from each population.
- Gather and record data from your samples.

Option 2:

- Read the question provided on the next page.
- Explain how the data samples might have been collected.

HANDOUT FOR COMPARING POPULATIONS

The data samples:

Sample Sentence Lengths

Sample Number	Harry Potter and the Sorcerer's Stone	The Absolutely True Diary of a Part-Time Indian
1	11	9
2	23	9
3	28	3
4	25	12
5	17	22
6	9	2
7	13	9
8	4	10
9	7	18
10	19	32
11	16	9
12	13	9
13	11	4
14	26	7
15	9	16

Sample Number	Harry Potter and the Sorcerer's Stone	The Absolutely True Diary of a Part-Time Indian
16	9	14
17	10	5
18	13	12
19	10	7
20	15	16
21	15	10
22	10	22
23	6	12
24	6	5
25	16	20
26	8	12
27	6	17
28	45	3
29	32	13
30	4	4

The question:

Which book has longer sentences (more words per sentence), *Harry Potter and the Sorcerer's Stone* by J. K. Rowling or *The Absolutely True Diary of a Part-Time Indian* by Sherman Alexie*?

The populations:

The populations are the collections of sentences in each book.

Note: I chose these two books because the authors use such different writing styles. I was curious to see if their different styles might show up in the lengths of their sentences.

CONVERSATION STARTERS FOR #1

What do you notice? What do you wonder?

For students who are creating their own questions:

I wonder what makes a good question for this problem?

Your question should (1) have a whole number answer, (2) be interesting to you, (3) be something for which it is practical to gather data, and (4) be about a large population (so that you need to select a sample).

I wonder if it matters how I phrase my question?

You should state the question clearly and carefully so that it refers to the population from which you are drawing the sample. For example, if you are selecting a sample from the students at your school, your question should not refer to all students in general.

For all students:

I wonder how large my sample should be?

Larger samples are generally better than smaller ones, although you eventually reach a point of diminishing returns. A sample size of 30 or 40 is typically sufficient for most populations assuming you use good sampling methods. (Interestingly, sample sizes greater than 10% of the population can create complications when applying more advanced statistical methods, because some members of the population are likely to be selected more than once.)

I wonder how I can be confident that my sample is representative?

The best way to ensure a representative sample is generally to select it *randomly*. This means that every member of the population has an equal chance of being selected.

I wonder what I can do if it is not practical to select a random sample?

Do the best you can. For example, in the solutions for Option 2, I chose a random page, then a random sentence on that page. *Systematic samples* may be a good alternative. (See the Introduction to this exploration.) Either way, when you present your results, describe the method you used, and discuss how your method may affect your results and conclusions.

For students using the question about the sentence lengths:

I wonder what it means to say that a book has "longer sentences" given that the lengths of the sentences vary so much?

"Longer sentences" may refer to the mean or the median number of words per sentence. You may want to calculate both at first. (Later in the exploration, we will focus on the mean length.) When you state your conclusions, you should be clear about how you chose to measure it.

I notice that using a systematic sample might be practical.

SOLUTIONS FOR #1

These solutions show results from the question about sentence lengths on the Handout for Comparing Populations (see page 93). Even if your students design their own experiments, the solutions here should give you some general ideas about what to look for in their work.

Selecting the samples: Because it was not practical to count the number of words in every sentence, I selected a sample of 30 sentences from each book.

Ideally, I would have selected a random sample from each book, but this was not practical for me. It would have involved counting the number of sentences in each book, choosing a random sentence number, finding the sentence, and counting the number of words. (It would have taken a long time to count the number of sentences in the entire book and to find, say, the 3,274th sentence!) Instead, I used a random number generator (on http://www.random.org) to choose a random page, then a random sentence from that page.

This selection process is not quite random, because not every possible sample has exactly an equal chance of being selected. For example, the sentences on half-empty pages are more likely than others to be selected. In fact, any sentence on a page having fewer sentences is likely to be selected more often. Can you see why?

Some other things to think about and discuss: Can you think of other ways in which the process is not perfectly random? Do you think that it might bias the sample toward longer or shorter sentences? Do you think that the samples for one book will be biased differently than the other due to the sampling method? Can you think of better ways to select a representative sample? Does it make sense to use a systematic sample?

I wonder if would work to simply flip to a page in the book and point to a sentence?

People are notoriously poor at making random selections. Also, long sentences take up more room on the page and may be more likely to be pointed to than short ones.

When you collect data, you may need to make decisions about "what counts." For example, I made the following decisions: a sentence belongs to the page on which it starts; poems, titles, lists, etc. are not counted as sentences; hyphenated words always count as a single word; semicolons and dashes do not mark the beginning of a new sentence.

STAGE 2

In Problem #2, students use their knowledge of data displays and measures of center and variability to analyze and interpret the data from Stage 1. See the Appendix at the end of the book if you would like some help with these.

What You Will Need

» Graph paper and calculators

What Students Should Know

» Understand concepts from Exploration 2: A Day at the Races.
» Calculate mean absolute deviation (see the Appendix or Exploration 1: Playing With Data).

What Students Will Learn

» Use samples to make inferences about a population.

Problem #2

When you analyze the data from a sample, you can use the same methods that you have already learned to display and make sense of data.

Directions

- Create a histogram and box plot for the data from each sample.
- Describe the important features of the histograms and box plots.
- Calculate the mean and the mean absolute deviation for each sample. Make some observations and comparisons.
- Interpret your data.

Diving Deeper

Use different bin sizes to create one or two more histograms for one of the populations. Describe how the appearance of the histogram changes. Which bin size do you think is best? Why?

CONVERSATION STARTERS FOR #2

What do you notice? What do you wonder?

I notice that the means are greater than the medians for both data sets.

I wonder if I can predict the mean/median comparison by looking at the histograms?

Notice the tails on the right side of each graph. (The graphs are "right-skewed.") How do they affect the mean? The median?

I notice that the data have some outliers.

I wonder whether I should include the outliers when I calculate the mean and the mean absolute deviation?

Try it both ways and look at the results. Think about what the outliers tell you as you try to understand the answer to the original question.

I wonder how interpreting the data is different than describing the graphs?

Describing the graphs involves looking for features such as clusters, gaps, outliers, and symmetry or "skewness." Interpreting the data involves using the appearance of the graphs to tell what they mean—in other words, relating the data to the original question.

SOLUTIONS FOR #2

Histograms and box plots:

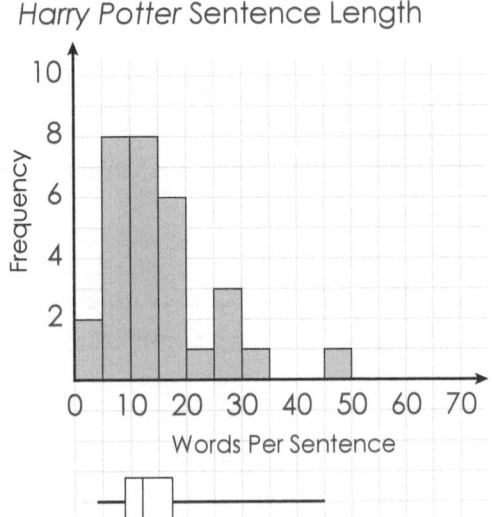

Minimum: 4; Lower Quartile: 9; Median: 12; Upper Quartile: 17; Maximum: 45

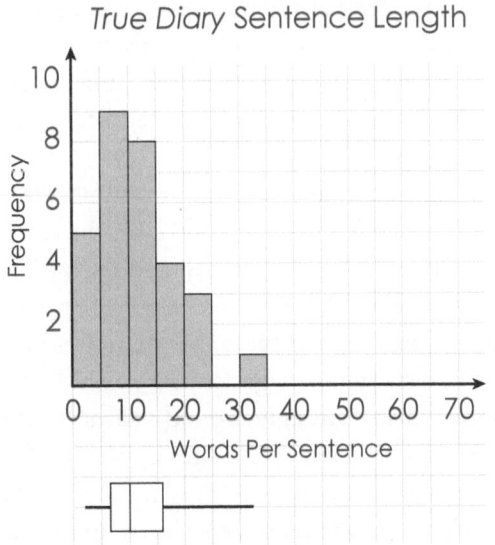

Minimum: 2; Lower Quartile: 7; Median: 10; Upper Quartile: 16; Maximum: 32

Description of the graphs: The bars on both graphs start moderately low, increase quickly, and then taper gradually to the right. The distributions are *right skewed* (due to the tails on the right). Both graphs also show a strong cluster of values between about 5 and 15 words per sentence. The graph for *Harry Potter* extends further to the right with a larger gap between the higher values, while the *True Diary* graph has somewhat more values on the far left.

Both box plots have a short whisker on the left and a long one on the right. The boxes are approximately the same width, although the one for *True Diary* is slightly wider. The box for *Harry Potter* is also slightly to the right of the box for *True Diary*. The median bars for both boxes sit closer to the left side of the box.

Mean and mean absolute deviation:

Harry Potter	Mean: 14.5	Mean Absolute Deviation: 6.8
True Diary	Mean: 11.4	Mean Absolute Deviation: 5.2

Without the outliers:

Harry Potter	Mean: 12.8	Mean Absolute Deviation: 5.2
True Diary	Mean: 10.7	Mean Absolute Deviation: 4.6

The *Harry Potter* data have two outliers: 32 and 45. The *True Diary* data have one: 32. Removing these numbers may be questionable because, although very long sentences are relatively rare, they may be an important part of the authors' style. However, I chose to look at the data both ways in order get a clearer sense of what the numbers were telling me. The mean for *Harry Potter* is still greater than for *True Diary*—but noticeably less so when you focus on typical sentences.

The sentences in *True Diary* also have less variability as measured by MAD than in *Harry Potter*. This appears to be due mainly to Rowling's occasional very long sentences, because the difference decreases substantially when these are removed. However, *True Diary* has a slightly greater interquartile range!

Interpretation: The mean, minimum, median, lower quartile, upper quartile, and maximum are all greater for *Harry Potter*. However, the differences are not dramatic. Removing the outliers reduces the mean and the variability for both books but does not affect this general relationship. Overall, it appears that the sentences in Rowling's book are probably a little bit longer than those in Alexie's.

A closer look at the histograms suggests what might be behind this. For both books, sentence lengths cluster strongly around values between about 5 and 15 words. In other words, the books look fairly similar in terms of the "typical length" sentences. However, Alexie's book seems to have more very short sentences, while Rowling's book has more very long ones. This shows up in the fact that her histogram is more strongly skewed to the right (which also explains why her mean is so much greater than her median and why removing the outliers affected her mean more strongly).

Interestingly, the evidence about the variability of the two samples is mixed. One measure (the MAD) suggests a greater variability for *Harry Potter*, while the other (the IQR) points to *True Diary*. Because the distinction is small, it is not too surprising, and it may result from the greater abundance of very long sentences in *Harry Potter*.

While reading the books, I felt that Alexie's sentences were clearly shorter (and that this affected the flow and feel of the book). The results of the investigation so far seem to support this observation, but they also suggest that the difference may be less than I suspected and that it may have more to do with "extreme" sentences than with more typical ones.

In fact, given how close the means are (especially when the outliers are discarded), I have to wonder if the mean sentence length in *Harry Potter* is actually greater! Perhaps the apparent difference between the means happened by chance because of the samples that I chose. Are the means different enough that I can be confident in my conclusion that *Harry Potter* has longer sentences? (You will begin to explore this question in Problem #3.)

STAGE 3

When you use samples to compare a value for two different populations, there is always some uncertainty involved. How great does the difference between the means of your samples have to be in order for you to be confident that it reflects a real difference in the populations?

Before beginning Problem #3, ask your students to think about this question as it relates to their results in Problem #2. Are the differences in the means great enough that they feel confident in their conclusions? If not, can they think of anything they could do to increase their confidence?

Statisticians deal with this uncertainty by thinking about what would happen if they collected a lot more samples (of the same size) from the population. Would the means of these samples vary a lot, or would they stay close to the ones in your original samples? Fortunately, there are ways to answer these questions without actually collecting all of these samples! Less fortunately, these ideas take a lot of time and more advanced mathematical knowledge to understand.

In this exploration, your students *will* collect more samples (or analyze samples that I have provided) in order to gain insight into this approach. The end result will be a better understanding of what the means of the original samples are actually telling you.

What You Will Need

- » Graph paper and calculators
- » Additional data sets for *Harry Potter* and *True Diary* (pages 112–115) (unless you are researching your own question)

What Students Should Know

- » Understand concepts from Stages 1 and 2 of this exploration.

What Students Will Learn

- » Compare variability between samples from the same population.
- » Graph sample mean distributions and understand their characteristics.
- » Use the sample mean distributions to estimate uncertainty.

Problem #3

You may wonder if your sample means are different enough to draw conclusions about which book has longer sentences.

Directions

- Collect 20 representative samples of the same size as your original sample from each population (or use the samples provided for you).
- Calculate the mean of all 20 samples for each population.
- Create a histogram of the sample means for each population.
- Compare the two histograms, and interpret your observations.
- Compare your histograms to the original two histograms in Problem #2.

CONVERSATION STARTERS FOR #3

What do you notice? What do you wonder?

I notice that the sample means have a relatively small range.

I notice that it is easier to compare the histograms if they have the same size and scale.

> It may also help to align one histogram directly above the other. (See the Solutions for #3.)

I notice something different about the variability of these histograms compared to the ones in Problem #2.

I wonder what causes this difference?

I wonder what the means and medians are for the distributions in this problem?

> If you calculate them, you will notice that the mean of each histogram is nearly equal to its median. This tells you something about the symmetry of each distribution. (It may seem strange at first to calculate the mean or median of a bunch of means, but it can be very useful!)

I wonder if the mean of the 20 sample means is equal to the mean of all of the measurements?

> Yes, it is—but only because every sample contains the same number of sentences.

SOLUTIONS FOR #3

The samples: Students who are using the question about sentence lengths may use the data at the end of this exploration. They contain sentence lengths for a large collection of samples from both books. Of course, if they have copies of the books, students may even collect their own samples!

Means of the samples: The set of data mentioned above contains 20 samples of 30 sentences each for both books. The means in the following tables are taken from these data. (If the task of calculating all of these means becomes too tedious for your students, ask them to share the work. Or, once they have calculated a few, you may simply give them the rest of the means.)

Harry Potter and the Sorcerer's Stone

Sample Number	Mean	Sample Number	Mean
1	14.53	11	12.00
2	10.93	12	12.37
3	11.97	13	10.80
4	13.03	14	10.90
5	12.97	15	11.40
6	15.10	16	13.43
7	14.30	17	13.93
8	13.10	18	13.27
9	11.87	19	8.33
10	14.43	20	16.90

The Absolutely True Diary of a Part-Time Indian

Sample Number	Mean	Sample Number	Mean
1	11.77	11	9.30
2	8.53	12	7.90
3	10.57	13	10.07
4	11.27	14	9.27
5	9.00	15	12.10
6	9.57	16	10.47
7	10.37	17	9.53
8	10.53	18	9.77
9	8.50	19	9.63
10	9.07	20	10.03

Histograms:

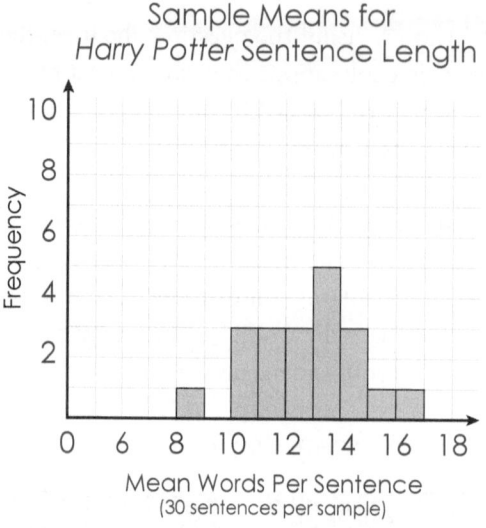

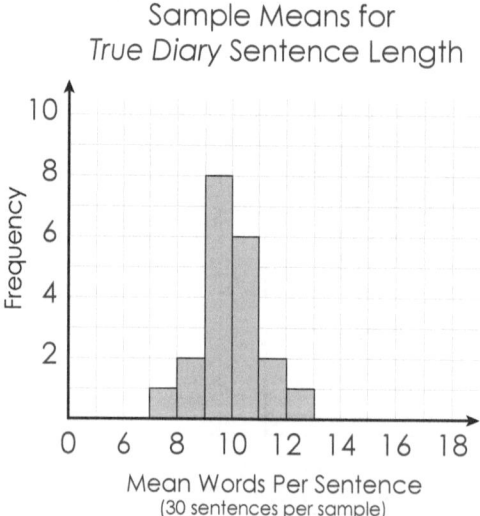

Comparing the two histograms: The center of the distribution for *Harry Potter* is clearly to the right of the one for *True Diary*, suggesting that a typical sentence in *Harry Potter* is in fact longer than one from *True Diary*. However, there is an overlap between the two distributions, which means that there is some doubt about this conclusion!

The sample means for *True Diary* are less variable than for *Harry Potter*. This suggests that we probably know the mean for *True Diary* a little more precisely than for *Harry Potter*.

Comparing these histograms to the ones from Problem #2: The values near the centers of the histograms in this problem (which seem to be near about 13 and 9.8) appear to be less than the means calculated in Problem #2. Thus, the means from Problem #2 were probably both a little greater than the actual means.

These histograms have much less variability than the original histograms from Problem #2. This makes sense, because the length of any one sentence may be far from the mean length, but the means of the samples should be fairly close to the actual mean. (After all, the purpose of using the samples is to approximate the length of a typical sentence!)

Unlike the original histograms, the histograms of the sample means appear to be approximately symmetrical. In other words, the mean of any given sample is about equally likely to be greater than or less than the actual population mean. To further support the observation that both histograms are nearly symmetrical, try calculating the mean and median of the sample means for each book. You will notice that they are approximately equal.

Problem #4

The size of a sample may be important to consider when interpreting data.

Directions

- Suppose that you had used smaller samples to calculate the sample means in Problem #3. Describe how this might have affected your histogram. Explain your thinking.
- Suppose that you had used larger samples to calculate the sample means in Problem #3. Describe how this might have affected your histogram. Explain your thinking.
- Imagine you had a friend who insisted that the sentences in *True Diary* are longer. Use the histograms of the sample means from Problem #3 to evaluate your friend's claim and respond to it.

CONVERSATION STARTERS FOR #4

What do you notice? What do you wonder?

I notice that the means of small samples are sometimes more extreme than means of large ones.

I notice that the two histograms overlap, which suggests that some of *True Diary's* sample means may be greater than some of *Harry Potter's* sample means.

I wonder what would happen to the histograms of the sample means if I kept using larger and larger samples?

I wonder what would happen if I made histograms of the sample *medians*?

I wonder if it is possible to come up with a percentage probability that *Harry Potter's* sentences are longer?

This is the type of question that you explore in more advanced statistics courses.

SOLUTIONS FOR #4

The effects of using smaller samples: When you use smaller samples, the center of the histogram of sample means may shift somewhat, but the main effect will be to increase its variability. Thus, the histogram will probably be wider and (if you use the same number of samples as before) flatter. Assuming that your samples are not *too* small, it will also probably remain fairly symmetrical.

As an illustration, I used the *True Diary* data at the end of the exploration to find sample means for sets of 15 measurements instead of 30. (To make it easy, I used columns of 15 in the lists.) I found the following 20 sample means:

6.87, 13.27, 9.87, 12.07, 9.73, 12.13, 9.93, 10.13, 10.46, 7.80
10.13, 9.33, 11.33, 9.80, 8.60, 8.73, 10.53, 11.73, 7.73, 8.13

The resulting histogram is still quite symmetrical but is wider (has greater variability) than the histogram created using 30 sentences per sample.

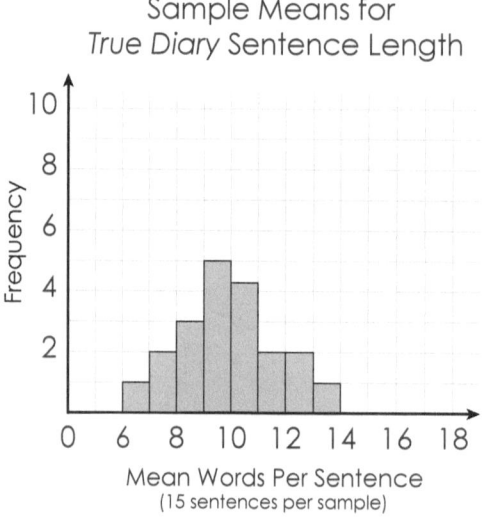

The reason for the increased variability is that when you have fewer numbers involved in calculating the mean, very large values may not get balanced as often by smaller values (or vice versa).

The effects of using larger samples: Again, the center of the histogram may shift somewhat, but now the main effect will be to *decrease* its variability. That is, the histogram will probably be narrower and (if you use the same number of samples as before) steeper. It will also probably remain fairly symmetrical.

Evaluating your friend's claim: It is possible but unlikely that your friend is correct. The regions where it is possible for *Harry Potter*'s sample means to be less than *True Diary*'s sample means occur where the histograms overlap:

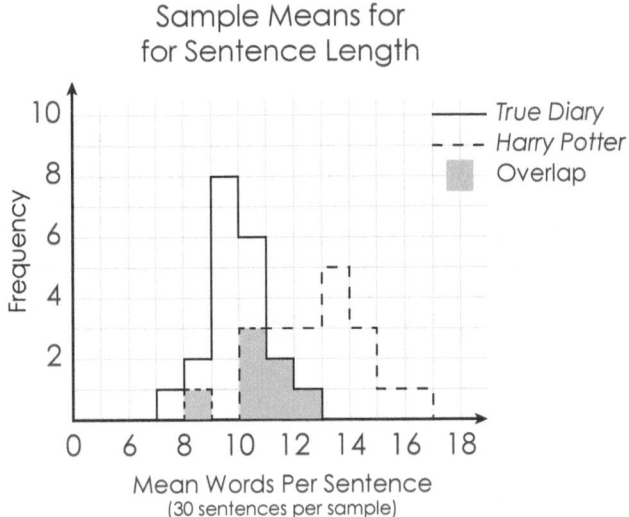

In other words, the amount of overlap between the histograms of the sample means tells us something about the likelihood that *Harry Potter*'s sentences are truly longer. To understand this better, imagine that there was no overlap. Then the true mean sentence length for *Harry Potter* would almost certainly be greater, because all of *Harry Potter*'s sample means would be greater than all of *True Diary*'s. Where the histograms *do* overlap, the sample mean for *Harry Potter* belongs to the left side of its histogram, and the sample mean for *True Diary* lies near the right side of its histogram, and in this situation, *True Diary*'s sample mean may be greater! However, because the region of overlap is relatively small, we can be quite confident that *Harry Potter*'s true mean sentence length is greater.

Some final notes: Some students may be curious to know exactly *how* confident we are in our conclusion. Answering this type of question is one of the main tasks of a high school statistics course. However, this need not prevent your students from doing some creative exploration of their own. For example, they might notice that you can form $20 \cdot 20 = 400$ possible pairs from the 20 sample means for each book. Of these, there are exactly 34 in which *True Diary*'s sample mean is greater than *Harry Potter*'s. This represents 8.5% of the pairings. Does this actually mean that you are $100\% - 8.5\% = 91.5\%$ confident that the sentences in *Harry Potter* are longer? Probably not. There are plenty of reasons to question an estimate like this. In fact, you would have to think carefully about what the phrase "91.5% confident" might even mean. However, the experience of playing with and reasoning about sample means is certainly valuable for students if they do not overinterpret their results!

Harry Potter and the Sorcerer's Stone

Number of Words per Sentence—10 Samples of 30 Sentences Each

11	23	28	9	10	13	25	17	9	10
15	15	13	4	7	10	6	6	19	16
13	16	8	6	11	26	9	45	32	4

27	5	9	7	9	13	10	30	19	2
1	4	4	9	14	19	14	16	13	3
8	19	4	5	18	10	7	14	14	1

15	7	10	27	3	7	6	1	17	14
24	17	24	13	6	27	6	14	7	20
2	6	16	14	8	5	16	10	2	15

10	7	21	6	6	16	14	52	18	19
5	7	7	4	21	11	6	4	30	6
21	16	6	22	7	2	19	3	10	15

15	26	10	7	32	18	8	10	15	4
6	12	25	19	7	6	2	18	11	8
5	23	5	7	15	28	6	10	17	14

32	15	10	8	39	12	4	13	26	3
19	15	10	9	25	8	24	14	11	9
19	14	15	13	18	16	10	29	8	5

12	3	6	27	11	1	31	4	5	5
2	7	5	7	21	65	44	28	2	29
4	31	14	6	23	3	1	7	17	8

11	31	6	4	5	7	18	17	33	15
1	3	18	20	7	61	16	5	26	22
2	8	13	3	8	4	5	5	16	3

6	25	10	13	25	1	8	17	2	4
4	42	26	8	5	18	13	3	10	3
17	18	11	4	14	9	10	13	8	9

12	8	5	15	9	15	12	14	17	13
8	36	9	20	27	46	7	16	15	2
14	8	15	13	7	8	11	9	11	31

Harry Potter and the Sorcerer's Stone

Number of Words per Sentence—10 Samples of 30 Sentences Each

18	18	9	17	6	5	15	10	14	2
2	3	11	6	3	12	10	25	10	10
2	33	6	8	3	6	19	14	46	17

12	9	18	6	13	5	13	12	6	7
9	12	3	8	13	17	3	8	36	5
36	3	5	20	13	28	7	14	13	17

15	2	5	18	13	11	8	10	4	22
12	5	9	6	4	8	10	12	18	6
5	9	12	15	11	15	7	20	19	13

8	3	18	14	32	6	15	9	10	6
5	3	12	21	12	7	7	28	2	8
16	2	8	20	5	15	4	13	3	15

2	9	8	5	11	4	12	4	6	25
7	14	21	14	28	3	8	8	19	11
27	10	7	10	4	3	25	6	16	15

8	6	9	41	14	23	6	4	3	12
10	19	6	10	37	24	34	7	33	9
12	13	1	5	12	12	3	14	3	13

18	11	21	3	12	11	5	15	3	23
11	12	8	7	20	15	27	17	14	8
5	23	3	18	15	5	48	18	20	2

9	6	3	9	33	2	22	17	10	17
5	6	6	10	6	39	22	1	3	50
23	11	1	6	10	2	5	30	29	5

9	12	9	19	14	4	18	17	4	2
3	4	3	8	7	14	2	2	3	5
4	8	5	2	7	13	5	10	29	8

12	7	13	6	17	17	13	2	7	20
38	5	18	19	17	7	4	25	39	11
24	22	16	33	41	13	23	16	18	4

The Absolutely True Diary of a Part-Time Indian

Number of Words per Sentence—10 Samples of 30 Sentences Each

9	12	9	32	4	19	22	10	9	7
3	2	18	9	16	14	7	22	20	3
5	16	12	12	13	12	10	5	17	4

9	31	5	10	3	8	3	9	9	8
8	7	5	16	7	20	12	18	10	5
7	7	12	3	3	3	8	4	4	2

21	12	4	17	29	9	24	9	13	6
13	17	12	6	6	23	11	6	7	6
4	5	9	11	6	4	7	5	4	11

4	5	4	18	2	13	6	6	19	6
1	35	7	7	5	20	7	7	12	20
11	15	15	2	18	13	9	22	10	19

3	16	4	26	13	1	2	10	11	9
1	15	22	7	8	18	14	4	8	6
4	4	10	5	13	5	7	15	4	5

10	6	4	9	3	8	9	8	10	9
7	8	7	29	7	12	6	12	7	11
18	30	5	3	10	10	9	3	8	9

6	8	7	10	17	9	25	18	7	6
7	15	9	10	5	1	12	9	2	21
6	7	7	7	3	10	11	29	16	11

25	17	7	17	21	17	4	3	8	1
4	3	18	10	20	6	24	13	3	3
19	7	11	5	4	4	8	11	15	8

11	1	12	9	8	8	5	6	2	17
2	8	17	10	6	21	7	7	8	4
9	13	6	9	11	5	13	12	2	6

6	2	11	5	4	3	1	10	1	7
13	2	41	7	6	6	8	28	21	5
9	13	8	7	4	11	16	7	6	4

The Absolutely True Diary of a Part-Time Indian

Number of Words per Sentence—10 Samples of 30 Sentences Each

15	21	28	24	5	6	8	14	13	7
3	14	5	9	10	4	3	8	1	7
11	6	9	9	11	5	2	4	4	13

7	5	8	3	10	4	7	13	3	12
9	2	4	4	6	4	11	6	6	7
11	11	6	7	9	15	8	18	18	3

20	10	5	3	28	14	6	9	9	1
6	14	9	5	8	1	5	7	14	11
3	9	21	16	10	7	18	4	21	8

4	3	4	14	18	7	5	18	11	9
22	1	8	6	4	12	1	8	8	5
10	29	10	1	5	7	5	23	14	6

14	4	1	7	21	19	17	5	17	37
9	3	6	1	6	15	6	26	5	9
11	12	7	16	6	40	4	18	5	16

29	15	16	2	17	2	6	8	4	4
8	11	16	8	3	12	8	29	8	4
5	11	10	5	13	5	1	11	34	9

13	10	10	4	7	10	6	11	6	13
9	9	9	4	10	5	5	4	5	4
29	6	16	4	14	3	20	12	22	6

3	28	3	14	23	5	19	6	22	6
13	10	8	8	6	7	6	4	18	3
7	8	4	18	14	9	4	7	1	9

8	5	2	4	4	4	9	23	23	13
9	4	12	14	5	15	4	14	19	5
1	6	14	18	3	12	6	10	11	12

12	9	19	1	14	16	11	8	11	7
7	6	6	4	6	17	23	15	18	3
4	6	10	11	10	6	8	8	8	17

Exploration 5

One More Time!

In this exploration, students learn more about situations in which actions such as flipping coins and spinning spinners are repeated. They discover how to use their knowledge of the individual probabilities of events to calculate probabilities for combinations of events.

Mathematicians have developed notations to describe these situations more easily. The "probability of event A" is often written $P(A)$. In this exploration, students will focus their attention on $P(A \text{ or } B)$ and $P(A \text{ and } B)$. As the problems become more complex, they will progress from using detailed tables and diagrams to organize their thinking toward more efficient approaches grounded in the new ideas they have discovered. However, please do not rush this process. When your students are ready, they will usually begin to choose the faster approaches themselves.

DOI: 10.4324/9781003232780-8

STAGE 1

In Stage 1, your students will explore probabilities of combined events such as $P(A$ *or* $B)$ and $P(A$ *and* $B)$. Problem #1 focuses on $P(A$ *or* $B)$, the probability that either event A or event B will occur. They will discover that it is important to pay attention to whether the two events have outcomes in common.

Problem #2 deals with $P(A$ *and* $B)$, the probability that both events A and B will occur. For example, you could write the probability of "two consecutive heads" as $P(H$ and $H)$. What makes Problem #2 different is that, unlike the case of heads and tails, the individual outcomes are not equally likely. The end goal is for students to discover and make sense of the rule for multiplying probabilities. Encourage them to use their own ideas for writing down and organizing their outcomes. This creates great opportunities to discuss the advantages and disadvantages of lists, tables, tree diagrams, and other representations.

In Problem #3, students generalize their discoveries from Problem #2 to probabilities involving more than two events. They search for patterns, make predictions, and apply what they learn to answer new questions.

What Students Should Know

- » Write fractions as percentages.
- » Understand the meaning of and know procedures for multiplying fractions.
- » Create simple Venn diagrams.
- » Understand the terms *event* and *sample space* (see the introduction to the "Simulation Station" activity).
- » Find probabilities for simple events that consist of equally likely outcomes.

What Students Will Learn

- » Understand how $P(A$ *or* $B)$ is affected when A and B have outcomes in common.
- » Use lists, tables, and tree diagrams to record and organize sets of outcomes.
- » Develop and apply strategies for calculating probabilities when individual outcomes are not equally likely.
- » Discover the multiplication rule for $P(A$ *and* $B)$. Apply it, and understand why it works.

NAME: _____ DATE: _____

Problem #1

Sean writes down the following equations for rolling a standard six-sided die.

$$P(\text{prime}) = \frac{3}{6} \qquad P(\text{even}) = \frac{3}{6} \qquad P(\text{prime or even}) = \frac{3}{6} + \frac{3}{6} = 1$$

Directions

- Fill in the Venn diagram for this situation.

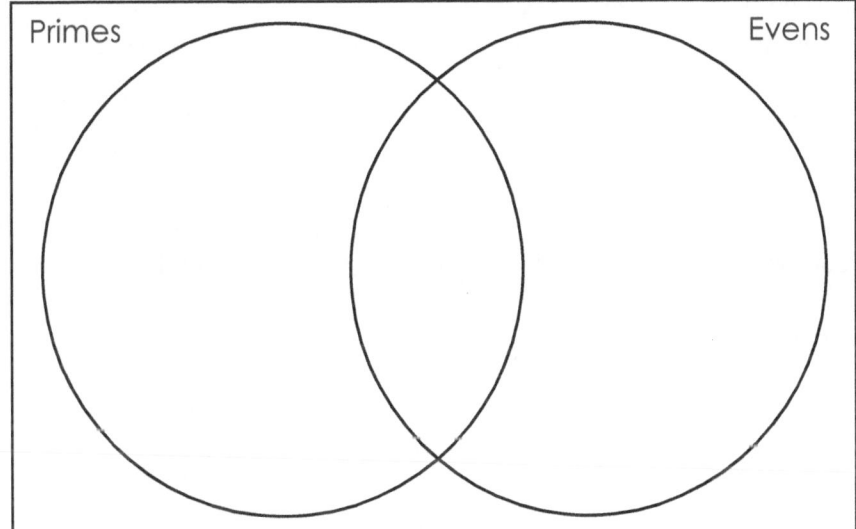

- Decide if Sean's equations are correct. Use your Venn diagram to explain.
- Create a Venn diagram showing the events "black card" and "odd card" for a standard 52-card deck. Use it to calculate $P(\text{black or odd})$. Explain.

Diving Deeper

Create a formula for calculating the probability, $P(A \text{ or } B)$.

CONVERSATION STARTERS FOR #1

What do you notice? What do you wonder?

I notice that the number 2 is both even and prime.
> In fact, it is the only even prime number!

I wonder if 1 is a prime number?
> No, it is neither prime nor composite.

I notice that the rectangle around the circles makes a place to include numbers that are on the die but not in either set.
> This rectangle is called the *universe set*. It contains every outcome in the sample space.

I wonder what counts as an *odd* card?
> Allow students to decide. In the solutions, I count Ace, 3, 5, 7, and 9 as odd cards.

I notice that there are many outcomes in the overlap between the "black card" and "odd card" events.

I notice that if two events have no outcomes in common, I can find the probability that one or the other will occur by simply adding their probabilities.

Note: Events that have no outcomes in common are called *mutually exclusive events*.

SOLUTIONS FOR #1

A Venn diagram: The number 2 is in both sets, because it is both prime and even. The number 1 is not in either set, because it is neither prime nor even.

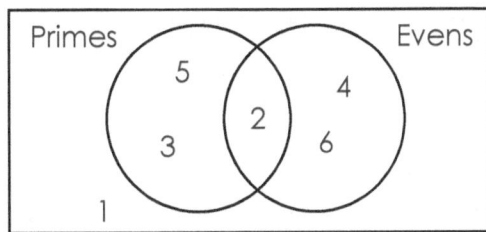

Sean's calculation: Sean's final equation is not correct. He calculates a probability of 1, which would mean that the outcome is certain. However, one of the numbers on the die ("1") is neither even nor prime, so the outcome cannot be certain. When he adds $\frac{3}{6} + \frac{3}{6}$, he counts the "2" twice. The correct probability is $\frac{5}{6}$.

Venn diagram for the cards:

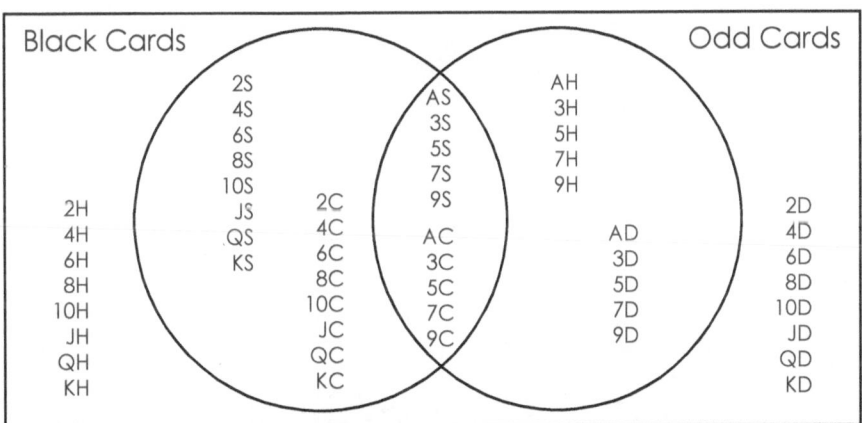

H, C, S, and D stand for hearts, clubs, spades, and diamonds respectively.

Calculation strategy 1: The number of outcomes in the event "black or odd" (that is, the numbers of outcomes listed inside the sets) is 36. The number of outcomes in the sample space (all of the cards) is 52. Because the individual outcomes are equally likely:

$$P(\text{black } or \text{ odd}) = \frac{36}{52} = \frac{9}{13} \approx 69\%$$

Calculation strategy 2: Add the probabilities for "black" and "odd" separately. Subtract the probability for cards that are black and odd, because they were counted twice in the sum.

$$P(\text{black } or \text{ odd}) =$$

$$P(\text{black}) + P(\text{odd}) - P(\text{black } and \text{ odd}) =$$

$$\frac{26}{52} + \frac{20}{52} - \frac{10}{52} = \frac{36}{52} = \frac{9}{13} \approx 69\%$$

Problem #2

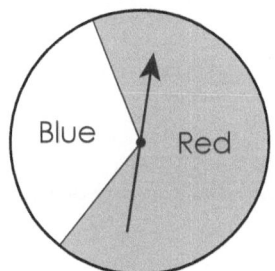

Jenna spins twice.

Directions

- Decide if it makes sense to use the formula:

$$P = \frac{\text{number of outcomes in the event}}{\text{number of outcomes in the sample space}}$$

  to calculate probabilities with this spinner. Explain your thinking.
- Determine the probability that Jenna will spin "red" both times. Use a table, tree diagram, or some other representation to justify your answer.

Diving Deeper

Determine the probability that Jenna will spin one of each color.

Testing the Waters

Determine the probability that Jenna will spin "blue" both times.

CONVERSATION STARTERS FOR #2

What do you notice? What do you wonder?

I notice that the outcomes "red" and "blue" are not equally likely.

I wonder what fraction of the circle each sector is?

The blue sector is $\frac{1}{3}$ of the circle. (Students may estimate or measure, or you may simply tell them.)

I notice that it does not make sense to add the probabilities for each red spin, because addition would increase my answer for the probability.

In fact, $\frac{2}{3} + \frac{2}{3} = 1\frac{1}{3}$, and the probability cannot be greater than 1!

I notice that the probability of spinning two reds in a row must be less than the probability of getting red on the first spin.

I wonder if it is possible to *create* equally likely outcomes.

SOLUTIONS FOR #2

Using the formula: Because there is 1 outcome in the event and there are 2 outcomes in the sample space, the formula would lead to a probability of $\frac{1}{2}$ for spinning "red," which is obviously incorrect. However, students may discover ways to use the formula by *creating* equally likely outcomes!

The probability of spinning two reds: The probability that both spins will be red is $\frac{4}{9}$.

Strategy 1: Because the outcomes are not equally likely, some students may divide the red sector into two equal parts. In order to keep track of the sectors, they may also label them separately using names such as "R1" and "R2."

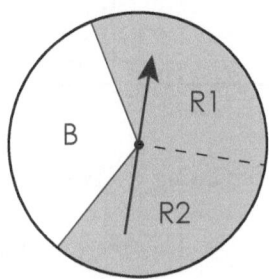

Now that there are equally likely outcomes for each spin, it makes sense to write down and count the outcomes for two spins. Students may use lists, tables, tree diagrams, or other representations to organize and guide their thinking. For example, a table might look like this:

B	B	B	R1	B	R2
R1	B	**R1**	**R1**	**R1**	**R2**
R2	B	**R2**	**R1**	**R2**	**R2**

Arranging outcomes in a table is a good way to ensure that you do not miss or double count any of them. Tables also highlight interesting patterns that you may not otherwise notice. This table shows 9 equally likely outcomes in the sample space, 4 of which (shown in bold) are in the event "two reds." This leads to a probability of $\frac{4}{9}$.

Some students may prefer to draw tree diagrams such as the one below.

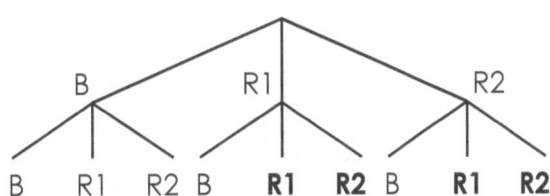

125

Each horizontal layer represents one spin, and each path from the top to the bottom shows a different outcome. Each of the nine labels at the bottom represents the end of a path. Again, four of the nine paths show the event "two reds."

Both the table and the tree diagram show how the 9 outcomes in the sample space result from calculating $3 \cdot 3$; there are three groups of the three outcomes. Similarly, the 4 outcomes in the event come from counting 2 groups of 2, or $2 \cdot 2$.

Strategy 2: Two thirds of the time, you expect to spin red, and $\frac{2}{3}$ of *this* time, you expect to spin red again. You may use area diagrams to represent the situation.

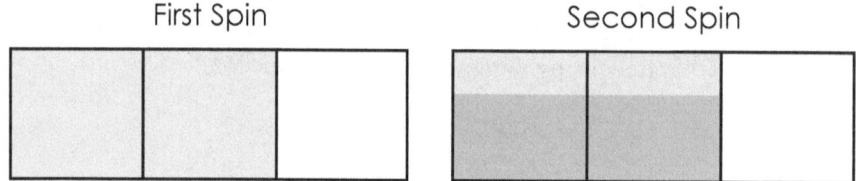

First Spin   Second Spin

Therefore, the probability of getting red both times is $\frac{2}{3}$ of $\frac{2}{3}$. You can see the answer of $\frac{4}{9}$ both by looking at the diagram and by calculating $\frac{2}{3} \cdot \frac{2}{3} = \frac{4}{9}$.

In Strategy 1, students are thinking of multiplying numerators and denominators separately, while in Strategy 2, they are thinking of $\frac{2}{3}$ as a single number. As they solve more problems, encourage them to think both ways.

Problem #3

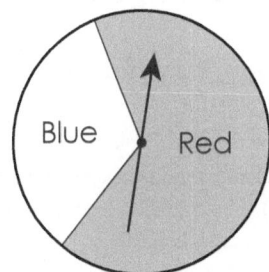

Jenna keeps spinning!

Directions

- Determine the probability for Jenna to spin three consecutive "reds."
- Determine the probability for Jenna to spin four consecutive "reds."
- Describe any patterns that you see, and explain what causes them.
- Find the minimum number of times Jenna must spin so that the probability of getting all "reds" is less than 10%.

CONVERSATION STARTERS FOR #3

What do you notice? What do you wonder?

I notice that each time Jenna spins, the probability of "all reds" decreases.

I wonder if I should continue making tables or tree diagrams?
 It is your choice. Tables, tree diagrams, and other representations guide your thinking and help you explain your reasoning. If you are able to think clearly and explain your ideas well without them, you may choose to use more efficient approaches.

I notice that I have a choice of counting outcomes (in the event and the sample space) or thinking of the probabilities directly (as fractions).

I notice that I can use patterns to find more efficient ways to figure out when the probability is less than 10%.

SOLUTIONS FOR #3

The probabilities for three reds and four reds are $\dfrac{8}{27}$ and $\dfrac{16}{81}$, respectively.

Strategies: Some students will continue making tree diagrams, tables, lists, or drawings. Encourage them to keep doing this as long as it helps them to understand and visualize the relationships. Ask them to share their approaches and discuss the advantages and disadvantages of each. Eventually, they will begin to notice patterns in their processes and results. Students who are able to predict these patterns immediately need not continue making diagrams as long as they can clearly explain *why* the patterns occur.

Patterns: The number of outcomes in the sample space multiplies by 3 each time you spin again. The number of outcomes in the event multiplies by 2. For example, if you spin three times, the sample space contains $3 \cdot 3 \cdot 3 = 3^3 = 27$ outcomes, while the number of events is $2 \cdot 2 \cdot 2 = 2^3 = 8$, resulting in a probability of $\dfrac{8}{27}$. Also, the probability becomes $\dfrac{2}{3}$ as great each time you spin.

In general, for n spins, the probability that all of the spins will be red is $\dfrac{2^n}{3^n}$ or $\left(\dfrac{2}{3}\right)^n$. The first expression results from using Strategy 1 from Problem #2. In this case, students are noticing "3 groups of 3 groups of 3," etc. The second expression comes from using Strategy 2, in which case they are thinking in terms of "$\dfrac{2}{3}$ of $\dfrac{2}{3}$ of $\dfrac{2}{3}$," and so on.

The number of spins needed to make the probability less than 10%: The probability becomes less than 10% on the sixth spin, because $\left(\dfrac{2}{3}\right)^6 \approx 0.088$ or 8.8 %, while $\left(\dfrac{2}{3}\right)^5 \approx 0.132$ or 13.2%.

STAGE 2

In Problem #4, students combine strands of knowledge from Stage 1 to solve a more complex problem. Be sure to let them choose strategies that suit their current understanding. Many will use tree diagrams or tables to organize and count outcomes, calculate probabilities, and explain their thinking. Others may have more efficient methods (especially if they have discovered formulas and understand why they work). As always, they will benefit from sharing and comparing strategies afterward.

A note on notation: I often write $P(A$ *and* $B)$ in a simpler form. For example, I show the probability of spinning two consecutive "reds" as $P(RR)$ instead of $P(R$ *and* $R)$. When the outcomes are numbers, I separate them, writing $P(5\ 7)$ instead of $P(57)$. This notation usually feels natural for students, and it makes complicated expressions easier to read. However, be sure that they remain aware of actual meaning of the expressions as they work with them.

What Students Should Know

» Understand concepts from Stage 1.

What Students Will Learn

» Solve complex problems involving both $P(A$ *and* $B)$ and $P(A$ *or* $B)$.

Problem #4

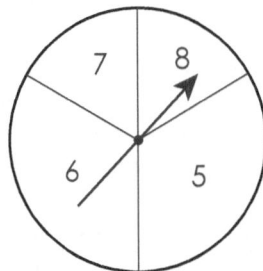

Gabriela likes prime numbers. Luana prefers Fibonacci numbers.

Directions

A friend spins the spinner twice and shows the two numbers to the girls.

- Find the probability that each girl will see a number that only she likes.
- Find the probability that at least one of the girls will like both numbers.
- Explain your thinking for both questions.

Diving Deeper

- Describe the *complement** of each of the two events above.
- What do you notice about the probability of each of the complements?

*Note: The *complement* of an event is another event that consists of all outcomes in the sample space that are *not* in the original event.

CONVERSATION STARTERS FOR #4

What do you notice? What do you wonder?

I wonder what a Fibonacci number is?

The Fibonacci numbers form an infinitely long list that begins 1, 1, 2, 3, 5, 8, 13, etc., so 1, 2, 3, 5, 8, and 13 are all Fibonacci numbers. These are the only ones you need to know for this problem, but you might be interested in finding a pattern to figure out how the list continues!

I wonder what the fractions for each sector of the spinner are?

"5" and "6" each cover $\frac{1}{3}$ of the circle. "7" and "8" cover $\frac{1}{6}$ of it. (Students may estimate or measure, or you may simply tell them.)

I notice that this problem combines ideas from Problem #1 and Problem #2.

I notice that I cannot use the formula for equally likely outcomes unless I create more sectors on the spinner.

I notice that it helps to make a Venn diagram showing which numbers each girl likes.

I notice that there are common outcomes involved in finding the probability that at least one of the girls will like both numbers.

SOLUTIONS FOR #4

The probability that each girl will see a number that only she likes is $\frac{1}{18}$. The probability that at least one girl will like both numbers is $\frac{7}{18}$.

Identifying the outcomes in the event for the first question: Many students will find it helpful to begin with a simple Venn diagram. L stands for numbers that Luana likes, and G represents numbers that Gabriela likes. (5 and 7 are prime numbers. 5 and 8 are Fibonacci numbers. 6 belongs to neither category.)

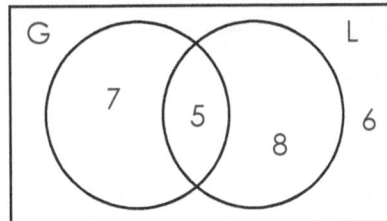

Only Gabriela likes the number 7, and only Luana likes 8. Therefore, the two numbers that were spun must include both 7 and 8. The two outcomes that apply are 8 7 and 7 8. Thus, the goal is to find $P(8\ 7\ or\ 7\ 8)$.

Finding the probability for the first question: Some students may calculate the probability of the outcome 7 8 as $\frac{1}{6} \cdot \frac{1}{6} = \frac{1}{36}$. The probability for 8 7 is the same. Therefore,

$$P(8\ 7\ or\ 7\ 8) = \frac{1}{36} + \frac{1}{36} = \frac{2}{36} = \frac{1}{18}$$

Many students will continue creating tables or tree diagrams to illustrate their ideas. (Even students who feel that they no longer need these visualizations in order to solve the problems may benefit from doing this occasionally. It helps ensure that their procedures remain grounded in the mathematical ideas.) They could begin by dividing the spinner into 6 congruent sectors and labeling them to distinguish the different 5s and 6s.

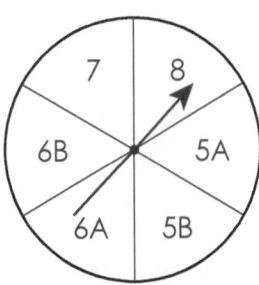

The outcomes in the sample space would then look like this:

5A 5A	5A 5B	5A 6A	5A 6B	5A 7	5A 8
5B 5A	5B 5B	5B 6A	5B 6B	5B 7	5B 8
6A 5A	6A 5B	6A 6A	6A 6B	6A 7	6A 8
6B 5A	6B 5B	6B 6A	6B 6B	6B 7	6B 8
7 5A	7 5B	7 6A	7 6B	7 7	7 8
8 5A	8 5B	8 6A	8 6B	8 7	8 8

Two of the 36 outcomes (the black highlights with light letters) are in the event. Thus, the probability is $\frac{2}{36}$.

Identifying the outcomes in the event for the second question: The number combinations for which Gabriela likes both numbers are 5 5, 7 7, 5 7, and 7 5. The combinations for which Luana likes both numbers are 5 8, 8 5, 8 8, and 5 5. These are shown in light highlights above.

Finding the probability for the second question: The table shows that 14 of the 36 outcomes are in the event, so the probability is $\frac{14}{36}$ or $\frac{7}{18}$. Again, some students may do the calculation without a table or tree diagram:

$$P(5\ 5\ or\ 7\ 7\ or\ 5\ 7\ or\ 7\ 5\ or\ 5\ 8\ or\ 8\ 5\ or\ 8\ 8)=$$

$$\frac{1}{3}\cdot\frac{1}{3}+\frac{1}{6}\cdot\frac{1}{6}+\frac{1}{3}\cdot\frac{1}{6}+\frac{1}{6}\cdot\frac{1}{3}+\frac{1}{3}\cdot\frac{1}{6}+\frac{1}{6}\cdot\frac{1}{3}+\frac{1}{6}\cdot\frac{1}{6}=\frac{14}{36}=\frac{7}{18}$$

As with Problem #1, there is some overlap between the events. Therefore, students need to be careful not to count outcomes twice. This Venn diagram shows the outcomes in the two events. (G and L now stand for number combinations in which Gabriela and Luana respectively like both numbers.)

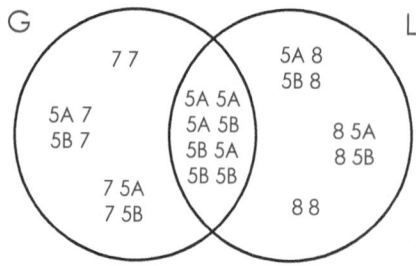

If you add the outcomes in Gabriela's set to those in Luana's set, you obtain 9 + 9 = 18. However, you will have counted the four that they have in common twice. Therefore, you must subtract 4 to obtain the correct number of outcomes: 18 − 4 = 14.

STAGE 3

Problem #5 presents students with a new type of challenge! Rather than calculating probabilities for a spinner, they work in reverse; they design their own spinner for a contest at a Math Family Fun Night. (This is similar to problems in which students try to adjust the rules of probability games to make them fair, but it is more challenging.) Many students will find it helpful to continue creating tables, lists, or tree diagrams, but the complexity of the problem may eventually encourage some of them to turn to more efficient methods.

In Problem #6, students apply their knowledge of data distributions to help them determine whether their spinner works as intended. This offers another opportunity for them to appreciate the importance of considering variability when analyzing data.

What You Will Need
» Materials for making a spinner (recommended): Poster board, green markers, compasses, protractors, etc., may be used for the circle. A pushpin and a paper clip work well for the spinning device.

What Students Should Know
» Use tables, lists, and tree diagrams to organize information about outcomes of probability experiments.
» Understand the multiplication rule for probabilities.

What Students Will Learn
» Apply the multiplication rule for probabilities to solve a challenging problem.
» Recognize and extend patterns in probability calculations.
» Consider variability when making decisions involving probabilities.

Problem #5

You are coordinating a Math Family Fun Night at your school. As each family enters, they will spin a spinner twice in an attempt to land on the school colors: green and white. Families who spin "green" and then "white" receive two gift certificates to local restaurants. Those who spin "white" and then "green" receive one gift certificate. Based on past experience, you expect about 150 families to attend. You have 40 gift certificates available to distribute.

Directions

- Design a spinner that will work well for this situation. Explain your thinking.
- Test your design by making the spinner and using it.

CONVERSATION STARTERS FOR #5

What do you notice? What do you wonder?

I notice that the probability of spinning white then green is equal to the probability of spinning green then white.

I notice that I expect to give away an average of 1.5 tickets per successful spin.

I notice that, no matter how I design the spinner, I may run out of gift certificates or have some left over.

I wonder what is the best way to measure the sizes of the sectors (fractions, percentages, or degrees)?

I wonder what is a reasonable first guess for the size of one of the sectors on the spinner?

I wonder if it matters which sector is which color?

I wonder whether it is better for the probability of spinning a green and a white to be a little too large or a little too small?

I wonder if I could create a workable spinner with more than two colors?

SOLUTIONS FOR #5

Option 1 for the spinner design: Create a green sector that covers slightly less than 10% of the circle and a white sector that covers the remaining portion. This is equivalent to approximately 35° and 325° (or about $\frac{1}{11}$ and $\frac{10}{11}$ of the circle) respectively. This will typically result in slightly fewer than 40 gift certificates being distributed.

Reversing the colors of the sectors will not affect the outcome (although it may affect the experience that families have as they spin). Because you cannot be sure what will happen (we are talking about probability after all!), you may choose to adjust the percentages by small amounts depending on whether you prefer to make it more likely to have gift certificates left over or to run out of them before all families have had a chance to spin.

Sample thinking process: Because the probability of spinning green followed by white is equal to the probability of spinning white followed by green, every two successful spins will typically result in giving out three gift certificates. Therefore, the number of successful spins should be about $\frac{2}{3}$ of 40, or approximately 26.7. Because $\frac{26.7}{150} = 0.178$, you want to create a spinner for which families will receive at least one gift certificate slightly little less than 18% of the time. In symbols:

$$P(\text{GW } or \text{ WG}) \approx 0.18$$

One way to decide on sizes for the sectors is to use a guess, test, and revise process. For example, if the spinner is green and white, then:

$$P(\text{GW } or \text{ WG}) = \frac{1}{3} \cdot \frac{2}{3} \cdot 2 = \frac{4}{9} \approx 44.4\%$$

This is much too high. However, it is interesting to continue investigating fourths, fifths, and so on. (*Note*: The multiplication by 2 is to account for both orders in which the colors may appear.)

$$\frac{1}{4} \cdot \frac{3}{4} \cdot 2 = \frac{6}{16} \approx 37.5\%$$

$$\frac{1}{5} \cdot \frac{4}{5} \cdot 2 = \frac{8}{25} \approx 32\%$$

$$\frac{1}{6} \cdot \frac{5}{6} \cdot 2 = \frac{10}{36} \approx 27.8\%$$

There are many patterns here. You can use them to predict a fraction of $\frac{18}{100}$ (18%) for $\frac{1}{10}$ and $\frac{9}{10}$. Because the desired probability is slightly less than 18%, you may choose to make the smaller sector somewhat less than $\frac{1}{10}$ of the circle.

Option 2 for the spinner design: Suppose that you want to create a spinner that has more than two colors. The simplest way to do this is probably to make the green and white sectors the same size, while the rest of the spinner has other color(s).

Suppose also that you decide to begin by keeping your calculations as precise as you can. The goal is for the number of successful spins to be $\frac{2}{3}$ of 40, which is equal to $\frac{80}{3}$. If you assume 150 spins, the desired fraction of successful spins is:

$$\frac{\frac{80}{3}}{150} = \frac{80}{3} \cdot \frac{1}{150} = \frac{8}{3} \cdot \frac{1}{15} = \frac{8}{45}$$

If the green and white sectors are to be the same size, then you are looking for a number that multiplied by itself equals $\frac{4}{45}$; in other words, $\sqrt{\frac{4}{45}}$. This represents approximately 0.29814 of the circle. Thus, the green and white sectors should each be about:

$$0.2981 \cdot 360° \approx 107.3°$$

This leaves about $360° - 2 \cdot 107.3° \approx 145.3°$ for the rest of the spinner, which may be any other color(s) you like!

The two options that I just described represent two extremes: no third sector (option 1) or the largest possible third sector (option 2). There is an infinite number of other solutions between these extremes! See the Algebra Connections section at the end of the exploration to read more about this.

Making and testing the spinner: Students should measure carefully when they create their spinners. Make sure they understand that the more times they perform the experiment, the more the likely it is that their results will be typical. When they test the spinners, they will need to strike a balance between practicality and spinning enough times to get informative results. Encourage students to compare the results of their experiments. Do small differences in the sizes of the sectors make a noticeable difference in their outcomes?

Problem #6

You have finished designing and creating your spinner for Math Family Fun Night. You have tested it by spinning many times.

Directions

- Analyze the data from the tests of your spinner*.
- Use your analysis to decide if you want to make changes to your spinner's design. Explain your thinking.

*Note: Alternatively, you could design and carry out a simulation with a random number generator and analyze the results of this instead. This will help you check that your calculations of the sizes of the sectors are correct. However, of course, it will not ensure that you have measured and drawn the angles accurately on your spinner.

CONVERSATION STARTERS FOR #6

What do you notice? What do you wonder?

I wonder whether it is better for the probability of spinning a green and a white to be a little too large or a little too small?

I wonder how much variability to expect in the number of gift certificates I would give out after 150 families spin?

I wonder what my analysis should include?

    It should almost certainly involve graphing the data. (Histograms and box plots would be helpful.) It should probably also include calculations of summary numbers for central tendency (mean, median, and mode) and variability (range, interquartile range, or mean absolute deviation).

I wonder how to compare histograms for different numbers of experiments?

    Try using *relative frequency* histograms in which the vertical axis represents the *percentage* of the total number of experiments.

I notice that the mean, median, and mode are close to each other.

SOLUTIONS FOR #6

Encourage your students to create histograms and box plots and to describe the shapes of the distributions. They should also calculate measures of central tendency (mean, median, and mode) and variation (range, interquartile range).

As an example, I have created computer simulations for a spinner with two sectors: $\frac{1}{11}$ green and $\frac{10}{11}$ white. The first simulation shows 1000 trials; the second one shows 100,000 trials!

One More Time

Number of Certificates Distributed	Frequency: 1,000 Experiments	Frequency: 100,000 Experiments	Number of Certificates Distributed	Frequency: 1,000 Experiments	Frequency: 100,000 Experiments
11	0	1	33	59	4807
12	0	3	34	43	5255
13	0	8	35	41	5439
14	0	8	36	48	5452
15	1	22	37	63	5535
16	0	37	38	53	5319
17	0	62	39	55	5250
18	3	111	40	45	4781
19	4	182	41	52	4584
20	1	241	42	45	4185
21	4	381	43	40	3694
22	5	551	44	29	3411
23	8	746	45	26	2862
24	10	1044	46	23	2528
25	15	1386	47	12	2078
26	21	1757	48	17	1705
27	18	2185	49	21	1435
28	34	2633	50	16	1163
29	33	3094	51	11	972
30	31	3701	52	14	729
31	40	4038	53	4	573
32	36	4477	54	4	437

Table: One More Time, continued

Number of Certificates Distributed	Frequency: 1,000 Experiments	Frequency: 100,000 Experiments	Number of Certificates Distributed	Frequency: 1,000 Experiments	Frequency: 100,000 Experiments
55	2	299	63	0	21
56	1	245	64	0	16
57	1	160	65	0	11
58	4	140	66	1	7
59	0	95	67	0	3
60	2	56	68	0	5
61	3	48	69	0	0
62	0	29	70	0	2

Some students may be interested in analyzing these data in addition to their own. *Relative frequency* histograms (which use percentages instead of frequency counts on the vertical axis) will make it easier to compare the two simulations. Both distributions are approximately symmetric. Students can verify this visually or by noticing that the means, medians, and modes are all close to or equal to the same number, 37. The 100,000-trial simulation has a somewhat "smoother" appearance, as there are fewer localized dips and peaks.

Deciding on modifications: Students may notice quite a bit of variability in the outcomes for their spinner. However (based on my data), after calculating the interquartile range, they should see that more than half of the time the 150 families would win between 32 and 42 gift certificates, which probably feels manageable.

At the same time, because it is also true that they cannot predict whether they will run out of gift certificates or have too many, they should be prepared for either situation. If they have a preference, they should check that their tests tend to produce the desired results. For example, the "elevenths" spinner that I simulated would work better for those who prefer a greater likelihood of having a few gift certificates left over. Notice that while it is practical to make adjustments to the spinner to choose between the two situations, it may not be practical to reduce the variability.

ALGEBRA CONNECTIONS

Using the language of sets, there is an algebraic formula for calculating $P(A \text{ or } B)$. It involves the concepts of *union* and *intersection* of sets. The *union* of two sets, written $A \cup B$, contains every object that is in one set or the other. You may think of it as combining the two sets, understanding that common elements occur only once in the union. For example, if A represents the set of prime numbers, $\{2, 3, 5\}$, and B stands for the set of even numbers, $\{2, 4, 6\}$, from the collection $\{1, 2, 3, 4, 5, 6\}$, then $A \cup B = \{2, 3, 4, 5, 6\}$. You can picture the union by shading in both sets.

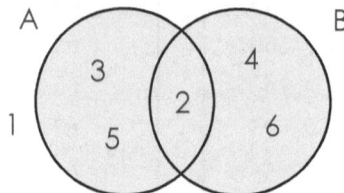

The *intersection* of two sets, written $A \cap B$, contains every object that is in both A and B. In other words, it contains the objects that are common to both sets. In the example above, $A \cap B = \{2\}$, because 2 is the only number that is contained in both sets; in other words, it is the only number that is both even and prime. You can picture the intersection by shading the overlap of the two sets.

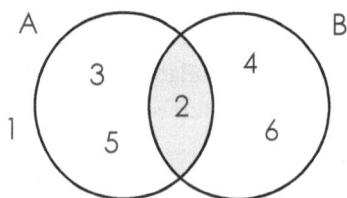

With this language, you can write an equation for $P(\text{prime or even})$ as:

$$P(A \cup B) = P(A) + P(B) - P(A \cap B) = \frac{3}{6} + \frac{3}{6} - \frac{1}{6} = \frac{5}{6}$$

The final subtraction undoes the double counting of the number "2" that occurs when you add the first two probabilities. Notice that if two events A and B are *mutually exclusive* (that is, they contain no common outcomes), then $P(A \cap B) = 0$, and the formula simplifies to:

$$P(A \cup B) = P(A) + P(B)$$

In Problem #5, students who know how to solve quadratic equations may be able to calculate the size of the sectors for which the expected number of successful spins is $\frac{2}{3}$ of 40. Suppose that *F* represents the fraction of the circle occupied by one of the two sectors, green or white. Assuming that there are only two colors on the spinner, the other sector has a measure of $1 - F$, and the probability that a family will win a certificate is $P(GW \text{ or } WG) = F(1-F) \cdot 2$. Now, $\frac{2}{3} \cdot 40 = \frac{80}{3}$, and $\frac{80}{3}$ is $\frac{8}{45}$ of 150.

Therefore, the quadratic equation for this situation is $F(1-F) \cdot 2 = \frac{8}{45}$, which has the solutions $\frac{45 \pm \sqrt{1305}}{90}$, or approximately 0.0986 and 0.9014. The sum of the two numbers is 1, as expected. And look at how well they agree with the results of $\frac{1}{11}$ and $\frac{10}{11}$ for Option 1! The smaller sector covers about 9.86% of the circle, or 35.5°, and the larger sector covers the rest.

What happens if you allow for a third sector that is neither green nor white? Suppose the fraction of the circle covered by the third sector is N. Then the quadratic equation becomes:

$$F(1-F-N) \cdot 2 = \frac{8}{45}$$

Solving for F using the quadratic formula and doing some simplifying and rewriting gives the result:

$$\frac{1}{2}(1-N) \pm \frac{\sqrt{225(1-N)^2 - 80}}{30}$$

You can learn a lot from studying this expression! First, notice that it represents an infinite number of solutions, one for each possible value of N. If $N = 0$ (in other words, if there is no third sector), then you get the earlier result for Option 1.

More generally, the expression $\frac{1}{2}(1-N)$ represents half of the sector that remains after the third sector is removed. Thus, if $\sqrt{225(1-N)^2 - 80}$ equals 0, both of the remaining sectors are the same size, $\frac{1}{2}(1-N)$! This is the situation from Option 2, and it corresponds to the greatest possible value of N.

To calculate this value, simply set the *discriminant* (the expression underneath the radical) equal to 0:

$$225(1-N)^2 - 80 = 0$$

Solving for N and choosing the value less than 1 gives $N = 1 - \sqrt{\frac{16}{45}} \approx 0.4037$.

Therefore, the greatest possible size of the third sector is approximately $0.4037 \cdot 360° \approx 145.3°$, and this occurs when the green and white sectors are the same size: $\frac{1}{2}(360 - 145.3) \approx 107.3°$ (the same result obtained earlier for Option 2)!

To summarize: Every value for the size of the third sector between $N = 0$ and $N \approx 0.4037$ gives a new spinner that solves the problem. When $N \approx 0.4037$, the green and white sectors are the same size (in order to get the greatest possible probability from the smallest total green/white sector size). As N decreases, the total fraction of the circle occupied by green and white increases, and the difference between the sizes of the green and white sectors becomes greater until $N = 0$, and the green and white colors occupy the entire circle. At this point, they attain the extreme values found in Option 1.

Exploration **6**

What Are the Chances?

In this exploration, students solve problems that reinforce, integrate, and extend concepts from earlier activities. The notion of *independent outcomes* (outcomes that do not affect other outcomes) emerges, and students discover and apply the idea of *conditional probability* (the probability that an event occurs given that another one has occurred previously). I do not place too much emphasis on this vocabulary when I teach the exploration, but I do make the students aware of the ways in which these problems are different than those they have been solving.

The problems in this exploration are fun to teach because of the variety of ways that students think about them. Their strategies fall into two broad categories—counting outcomes and multiplying probabilities. The former approach is concrete and less efficient, while the latter is generally more abstract and quicker to apply. The process of multiplying probabilities also brings students closer to the meaning of probability—long run relative frequencies of chance events (NGA & CCSSO, 2010)—because it enables them to focus on the probabilities themselves as opposed to simply listing outcomes and placing the results in a formula. However, both approaches offer opportunities for students to focus on the deeper meaning of the ideas.

The Solutions in this activity are longer than usual, because I show many levels in the typical progression in students' thinking. The key is for them to share their strategies and talk about why they are doing what they are doing, regardless of where they are in the progression. Ideally, most students will learn to move flexibly between the concrete and abstract approaches and be able explain how the two are connected. In order to reach this point, they may need to work on more problems than the ones provided here. Fortunately, it is fairly easy for them to create their own new problems. They may even test their results by carrying out simulations!

DOI: 10.4324/9781003232780-9

STAGE 1

In previous explorations, we quietly took for granted that the events were *independent*—that they did not affect probabilities of other events. When you use dice or spinners, it seems reasonable to assume that the result of one roll or spin has no effect on the next one! However, if you are drawing from a collection of cards or marbles, for example, and if you do not replace them before drawing again, the probabilities of future events change. In Stage 1, students begin exploring these types of situations and comparing them with more familiar ones.

The three problems in Stage 1 gradually increase in complexity in order to motivate students to develop more efficient strategies over time. However, although some students will begin using abstract and efficient strategies early on, others will never reach this point. If you allow them to choose their own strategies, they are likely to select those that best build on their current knowledge. By having your students share ideas and methods as they work, students who are using more elementary approaches can catch a glimpse of ideas that will eventually lead them to more sophisticated methods, and those who are already thinking more abstractly will be reminded of the foundational ideas that their strategies are built on.

Because students think in so many ways and at so many levels about these problems, the Solutions pages in Stage 1 are longer than usual. I have tried to include examples that show a range of approaches that I have seen in students. Don't worry if you do not understand all of the sample solutions right away. You will come to understand them better as you listen to your students.

What Students Should Know

>   Use tables, tree diagrams, or lists to record outcomes and to find $P(A \text{ and } B)$ for independent events A and B.

What Students Will Learn

>   Use tables, tree diagrams, lists, or formulas to record outcomes and to find $P(A \text{ and } B)$ when events A and B are dependent.

>   Develop efficient strategies for calculating probabilities, especially when it is not practical to list all outcomes in the sample space or event. Explain why these strategies make sense.

Problem #1

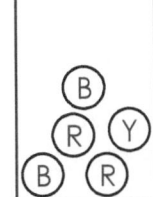

Shelagh has a bag with 2 blue, 1 yellow, and 2 red marbles. She mixes the marbles. Without looking, she draws one from the bag. She replaces the marble and repeats the process two more times.

Directions

- Calculate the probability that two of the marbles she draws are blue and one is red.
- Suppose that Shelagh does not replace the marble each time. Predict whether the probability will increase, decrease, or stay the same. Explain your thinking.
- Test your prediction by calculating the probability.

CONVERSATION STARTERS FOR #1

What do you notice? What do you wonder?

I notice that there will be a lot of outcomes.

I wonder what is the best way to keep track of all of the outcomes?
> Consider using a list, table, or tree diagram.

I wonder if I can find the number of outcomes in the sample space without showing all of them?
> A table or a tree diagram might have patterns that you can imagine extending without drawing the entire thing.

I notice that Shelagh can draw the colors in different orders.

I notice that it helps to give marbles different names when they are the same color.

I wonder if there is a way to reduce the number of branches in my tree diagram?
> It may be possible to combine some of the branches. How would you keep track of the original number of branches?

I wonder if there is a quick way to predict the number of outcomes in the sample space?

I notice that when Shelagh does not replace the marbles, it reduces the number of outcomes in the sample space.

I notice that when Shelagh does not replace the marbles, it also reduces the number of outcomes in the event by making it harder to draw a second blue marble.

I notice that sometimes it is impossible to draw certain colors on the second or third draws.

SOLUTIONS FOR #1

The probability with replacement is $\frac{24}{125}$. The probability without replacement is $\frac{1}{5}$.

Finding the probability with replacement: If each pair of blue and red marbles is labeled separately, a tree diagram for the first two draws begins like the following image.

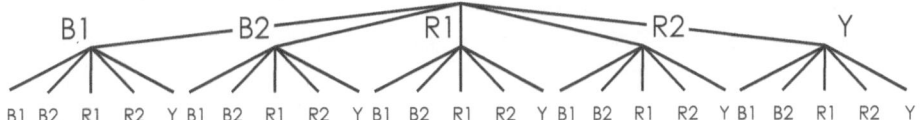

After two draws, there are 25 outcomes. It is difficult to continue the tree because of the large number of outcomes! However, it is easy to see that after the third draw, there would be 25 groups of 5 outcomes each for a total of 125 outcomes.

The complexity of the diagram may inspire you to look for more efficient methods. For example, you may combine branches if you keep a careful count of them!

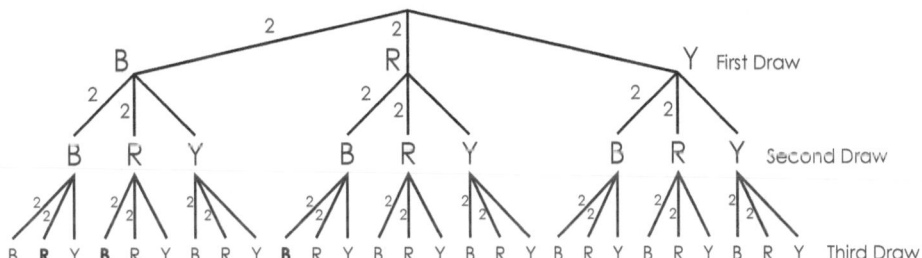

In this diagram, each pair of blue and red branches is combined into one branch and a "2" is placed next to it in order to show that the branch represents two outcomes. This reduces the number of branches in the final row from 125 to 27! Notice that you can determine the original number of branches simply by multiplying the numbers along any path. (Think of unmarked branches as having an unwritten "1.") For example, in the complete tree diagram, there would be $2 \cdot 2 \cdot 1 = 4$ BRY branches.

The three letters in bold show the ends of the paths BBR, BRB, and RBB in the event "two blues and a red." For each path, there are $2 \cdot 2 \cdot 2 = 8$ outcomes, which leads to a total of $8 \cdot 3 = 24$ outcomes for the event. Because there are 125 outcomes in the sample space, the probability of drawing 2 blues and 1 red if you replace each marble before drawing again is $\frac{24}{125}$ or 19.2%.

Predicting the effect of not replacing the marbles: This is difficult to predict, because not replacing the marbles reduces both the total number of outcomes and the number of outcomes in the event. Students' answers will vary.

Calculating the effect of not replacing the marbles: If you were to show the first two rows without combining branches, it would look like the following image.

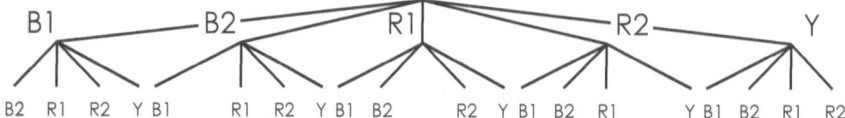

I have removed one branch from each part of the bottom row of the original tree diagram because if a marble is not replaced, it is not available to be drawn the second time! There are now 20 outcomes instead of 25. The final row would contain 3 new branches for each of these 20 for a total of $5 \cdot 4 \cdot 3 = 60$ branches. Not replacing the marbles decreases the total number of outcomes.

You may still combine branches in this case, but you need to be even more careful when counting them. I have shown the ones this time to make it easier to see how the number of branches changes each time.

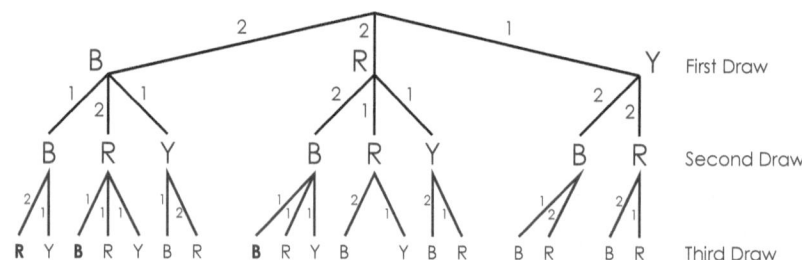

Once a color is used in a previous draw, the number of branches for that color decreases in the next row of the tree. In fact, branches disappear entirely when there are no longer any marbles of a certain color left!

The number of branches in the paths BBR, BRB, and RBB has changed. BBR now has $2 \cdot 2 \cdot 1 = 4$ paths; BRB has $2 \cdot 2 \cdot 1 = 4$ paths; and RBB has $2 \cdot 2 \cdot 1 = 4$ paths. This gives a total of $4 \cdot 3 = 12$ paths in the event. There are now 60 outcomes in the sample space. Therefore, the probability of drawing two blues and a red when the marbles are not replaced is $\dfrac{12}{60} = \dfrac{1}{5}$ or 20%. The probability has increased slightly from the case when the marbles were replaced.

Problem #2

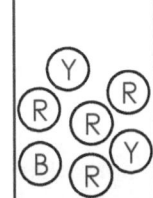

Anton mixes the marbles. Without looking, he draws one from the bag. He replaces the marble and repeats the process two more times.

Directions

- Calculate the probability that each marble is a different color.
- Make observations and describe patterns that you have seen in your work on Problems #1 and #2.

CONVERSATIONS STARTERS FOR #2

What do you notice? What do you wonder?

Look back at the Conversation Starters from Problem #1. Many of them still apply to this problem.

I notice that this problem is more complex than Problem #1.
There are more outcomes in the sample space, and Anton can draw the three colors in more ways.

I notice that the sum of the numbers on each "clump" of branches is equal to 7.
This is true when you combine branches. It makes sense, because there would have been seven branches in each clump if you had shown all of them.

I notice a pattern for the number of outcomes in the sample space.
Yes; the pattern involves an exponent, and you can discover it by analyzing your table or tree diagram.

I wonder if I can omit paths from my tree diagram when they are not part of the event?
Yes, but be sure that you still have a way to keep track of the number of outcomes in the sample space.

I wonder if there is a way to keep track of the probability for each branch as I draw it?
You can write the probability next to the branch instead of writing the number of branches that you combined.

I notice that the outcomes along each branch of a path are combined using "*and.*"
This is why you multiply their probabilities.

I notice that the events represented by the paths are combined using "or."
This is why you add their probabilities toward the end of a calculation.

I wonder if I can organize events and calculate probabilities without making diagrams?
A few students may be able to do this, but let them discover it for themselves—and ensure that they understand why their methods work.

I notice two ways of thinking about $P(A \text{ } and \text{ } B)$.
You can count outcomes or multiply probabilities.

I notice that the paths in my tree diagrams represent mutually exclusive events.
Yes. The events have no outcomes in common—usually because each event is a single outcome!

SOLUTIONS FOR #2

The probability is $\dfrac{48}{343}$.

Calculating the probability: As the complexity of the problems increases and students gain experience, their representations may gradually become less detailed and more abstract. In these solutions, I show a possible progression in their thinking. Do not press too hard for efficiency. Students should choose the level of detail and concreteness they need in order to make sense of the ideas.

Representation 1: Combine branches and number them for counting purposes.

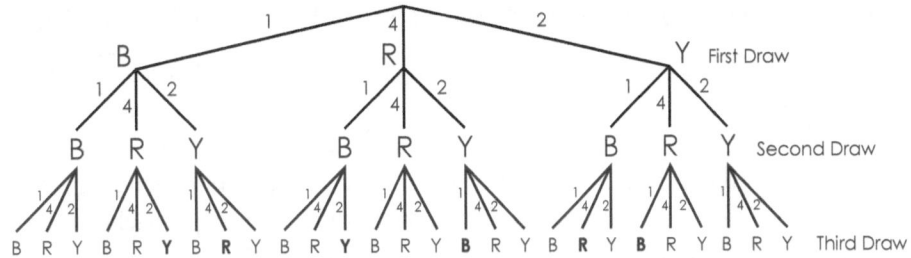

Representation 2: Combine branches, showing only those that correspond to the event.

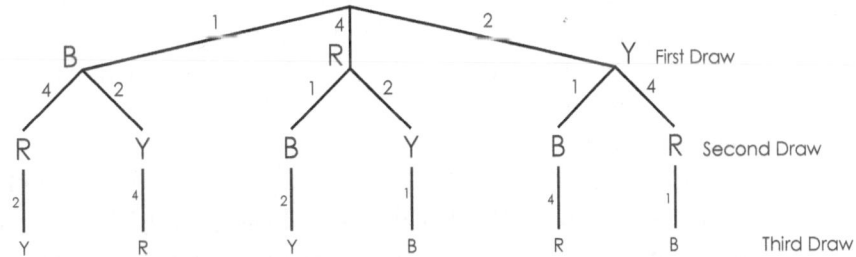

Representation 3: Show fractions for the probabilities on the branches.

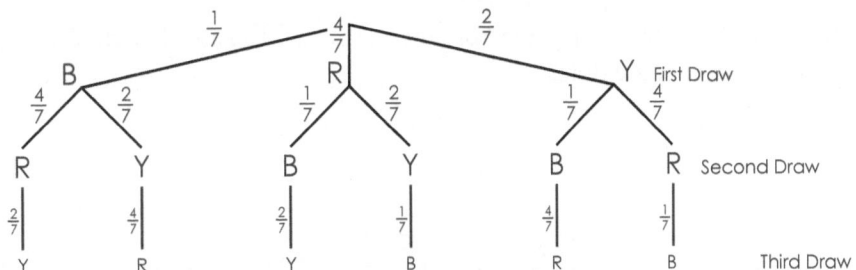

Representation 4: Make a list of events with their probabilities and calculations.

BRY: $\dfrac{1}{7} \cdot \dfrac{4}{7} \cdot \dfrac{2}{7} = \dfrac{8}{343}$ BYR: $\dfrac{1}{7} \cdot \dfrac{2}{7} \cdot \dfrac{4}{7} = \dfrac{8}{343}$

RBY: $\dfrac{4}{7} \cdot \dfrac{1}{7} \cdot \dfrac{2}{7} = \dfrac{8}{343}$ RYB: $\dfrac{4}{7} \cdot \dfrac{2}{7} \cdot \dfrac{1}{7} = \dfrac{8}{343}$

YBR: $\dfrac{2}{7} \cdot \dfrac{1}{7} \cdot \dfrac{4}{7} = \dfrac{8}{343}$ YRB: $\dfrac{2}{7} \cdot \dfrac{4}{7} \cdot \dfrac{1}{7} = \dfrac{8}{343}$

Results: The number of outcomes for any particular order of drawing the marbles is $1 \cdot 4 \cdot 2 = 8$. There are 6 orders in which they may be drawn (see Representation 4). The total number of outcomes in the event is $8 \cdot 6 = 48$. The number of outcomes in the sample space is $7 \cdot 7 \cdot 7 = 7^3 = 343$. Therefore, the probability is $\dfrac{48}{343}$, which is about 14%.

Observations: Students will reach their results in different ways depending on the representations they choose. The key difference between their approaches mirrors the two strategies discussed in the Solutions for Problem #2 from the exploration One More Time!: (1) counting outcomes and (2) multiplying probabilities.

The first two representations involve finding probabilities by counting outcomes, but they do not clearly show the number of outcomes in the sample space, so students may need to imagine what the tree diagram would like if all branches were drawn. There would be 7 branches for the first draw. On the second draw, there would be 7 groups of 7, or 49 branches. On the final draw, there would be 49 groups of 7, or 343 branches. Eventually, students will begin to notice that the total number of outcomes in the sample space is N^D, where N is the number of marbles and D is the number of draws.

Representations 3 and 4 involve thinking directly in terms of probabilities rather than individual outcomes. (The denominators do keep track of the number of outcomes in the sample space for you, however.) Students who use this approach may be thinking of the *meaning* of probability—the relative frequency of a chance event when you repeat a process many times. For example, an event of the type BRY will tend to occur $\dfrac{2}{7}$ of $\left(\dfrac{4}{7} \text{ of } \dfrac{1}{7}\right)$ of the time. However, it is also possible that these students are simply memorizing formulas. Ensure that they can give reasons for what they are doing.

Problem #3

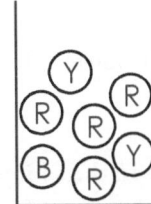

Anton draws three times, hoping to draw one of each color. However, this time he does not replace the marble after each draw.

Directions

- Predict whether the probability will be greater than, less than, or equal to the probability in Problem #2. Explain your thinking.
- Calculate the probability that each marble is a different color.

CONVERSATION STARTERS FOR #3

What do you notice? What do you wonder?

Look back at the Conversation Starters from Problems #1 and #2. Many of them still apply to this problem.

I notice that the number of outcomes in the sample space is less than in Problem #2.

I notice the number of branches coming from each outcome decreases by one on each draw (or it would if you drew all of them!).
    There would be seven branches for the first draw, six branches coming out of each of these, and five coming out of each of these.

I notice a new pattern for the number of outcomes in the sample space.
    There is one pattern when the marbles are replaced and a different pattern when they are not.

I notice that the number of outcomes in the event "one of each color" is the same as it was in Problem #2!

I wonder why this happens?
    Each color is different in each event.

I notice that it is easier to predict the effect of not replacing the marbles in this problem.

I notice that the order in which the colors are drawn does not affect the probability of the event.

SOLUTIONS FOR #3

The probability is $\frac{48}{210}$ (or $\frac{8}{35}$) which is greater than the probability in Problem #2.

Predicting the effect of not replacing the marbles: Some students may see that because every outcome consists of three different colors, the number of outcomes in the event will not change. However, the number of outcomes in the sample space will decrease, because there will be fewer marbles on the second and third draws. Therefore, the probability will be greater than it was in Problem #2.

Calculating the probability: I will show the same four types of representations as in Problem #2 so that you can compare and contrast them.

Representation 1: The number of marbles decreases each time you draw. For example, if you get a red marble on the first draw, there will be only 3 red marbles available on the second draw. No branch is drawn if there are no remaining marbles of a particular color.

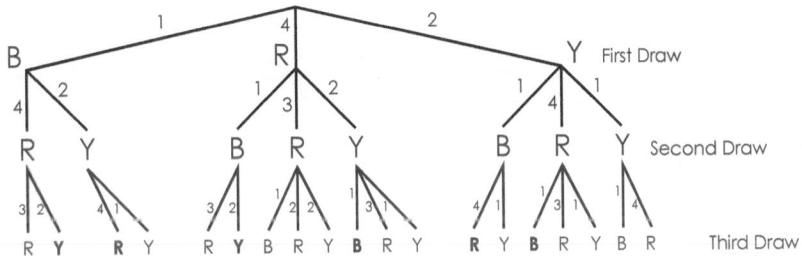

Representation 2: This representation looks the same as when the marbles were replaced. Because all three colors in the event are different, drawing a marble without replacing it does not affect the number of outcomes in any of the events of interest!

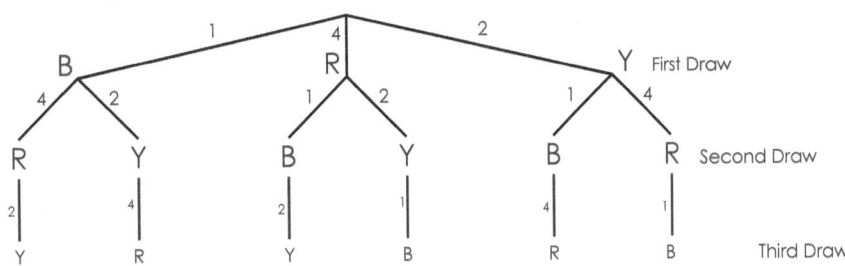

Representation 3: Only the denominators have changed from the case when the marbles were replaced. The total number of outcomes in the event is left unchanged for the same reason as above, while the number of marbles from which to choose decreases by one for each draw.

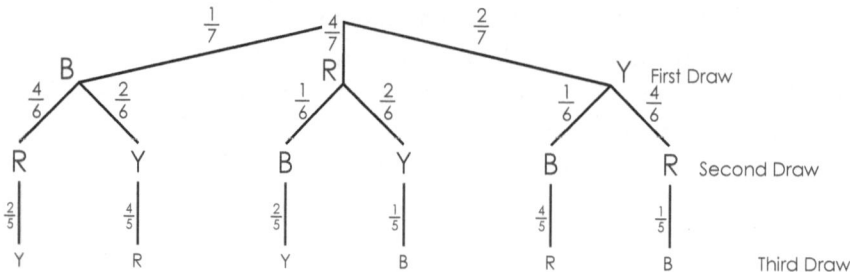

Representation 4: Again, only the denominators have changed, reducing the number of outcomes in the sample space. Students who use this representation must be careful to keep track of the number of marbles of each color for each drawing.

$$\text{BRY: } \frac{1}{7}\cdot\frac{4}{6}\cdot\frac{2}{5}=\frac{8}{210} \qquad \text{BYR: } \frac{1}{7}\cdot\frac{2}{6}\cdot\frac{4}{5}=\frac{8}{210}$$

$$\text{RBY: } \frac{4}{7}\cdot\frac{1}{6}\cdot\frac{2}{5}=\frac{8}{210} \qquad \text{RYB: } \frac{4}{7}\cdot\frac{2}{6}\cdot\frac{1}{5}=\frac{8}{210}$$

$$\text{YBR: } \frac{2}{7}\cdot\frac{1}{6}\cdot\frac{4}{5}=\frac{8}{210} \qquad \text{YRB: } \frac{2}{7}\cdot\frac{4}{6}\cdot\frac{1}{5}=\frac{8}{210}$$

The number of outcomes for any particular order of drawing the marbles is still $1\cdot4\cdot2=8$. Because there are 6 orders in which the colors may be drawn, the total number of outcomes in the event remains $8\cdot6=48$. The number of outcomes in the sample space has changed from $7\cdot7\cdot7=343$ to $7\cdot6\cdot5=210$. Therefore, the probability has increased from $\frac{48}{343}$ to $\frac{48}{210}$ (or $\frac{8}{35}$), which is approximately 23%.

STAGE 2

Problem #4 presents a situation involving probabilities related to outcomes and events that are neither independent nor mutually exclusive. I like to show students the tree diagram with no comment and see if they notice any issues with it before I prompt them for ideas. Allow your students to suggest ideas for questions to investigate. At the end of the discussion, you may use the directions as written or build in some of your students' suggestions.

What Students Should Know

» Find probabilities involving events that have outcomes in common.
» Find probabilities involving outcomes that are not independent.

What Students Will Learn

» Solve a problem in which the outcomes and events are neither mutually exclusive nor independent.

Problem #4

A: Ace
D: Diamond
N: Neither

You have a standard deck of 52 playing cards.

Directions

- Explain what might be happening in this probability experiment. Describe the difficulty with using the tree diagram above.

- You draw two times without replacing the first card. Calculate the probability that one card is an ace and the other is a diamond. Revise the tree diagram if necessary.

CONVERSATION STARTERS FOR #4

What do you notice? What do you wonder?

I notice that the tree diagram has two rows, which suggests two draws.

I notice that is not practical to draw every branch of the tree diagram.

I notice that the events "Ace" and "Diamond" are not mutually exclusive. (They share a common outcome.)

I wonder if it is possible to use the tree diagram as it is, even though the events are not all mutually exclusive?

I wonder how I can change the tree diagram to make all of the events mutually exclusive?

I wonder how it would affect the probability if I replaced the card after the first draw?

SOLUTIONS FOR #4

The difficulty with using the tree diagram: The tree diagram appears to be related to an experiment involving drawing two cards from a deck. The difficulty is that the events "Ace" and "Diamond" are not mutually exclusive (they overlap), because the Ace of Diamonds belongs to both events. One way to fix this problem is to list the Ace of Diamonds as a separate event.

A revised tree diagram:

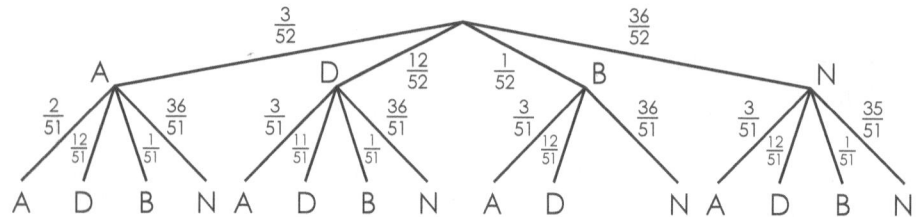

A: Ace (not diamond) B: Both Ace and Diamond
D: Diamond (not ace) N: Neither Ace nor Diamond

Probability: The outcomes in the event are AD, AB, DA, DB, BA, and BD:

AD: $\dfrac{3}{52} \cdot \dfrac{12}{51} = \dfrac{36}{2652}$ AB: $\dfrac{3}{52} \cdot \dfrac{1}{51} = \dfrac{3}{2652}$ DA: $\dfrac{12}{52} \cdot \dfrac{3}{51} = \dfrac{36}{2652}$

DB: $\dfrac{12}{52} \cdot \dfrac{1}{51} = \dfrac{12}{2652}$ BA: $\dfrac{1}{52} \cdot \dfrac{3}{51} = \dfrac{3}{2652}$ BD: $\dfrac{1}{52} \cdot \dfrac{12}{51} = \dfrac{12}{2652}$

$$\frac{36 \cdot 2}{2652} + \frac{12 \cdot 2}{2652} + \frac{3 \cdot 2}{2652} = \frac{102}{2652} = \frac{1}{26} \approx 3.8\%$$

Students' strategies will vary. For example, some may include only the branches in the tree diagram that show the relevant event.

STAGE 3

Problems #5 and #6, unlike most Stage 3 problems, are not necessarily more challenging than those in Stages 1 and 2. However, they do offer students an opportunity to synthesize some of what they have learned about simulations, multiplication of probabilities, and data representation and analysis. They also introduce students to the *geometric distribution*, a well-known probability distribution that has strong connections to geometric sequences and exponential functions, concepts that they will encounter in future algebra courses.

What You Will Need

>> Graph paper
>> A computer or a calculator that generates random numbers (optional)

What Students Should Know

>> Understand and apply the multiplication rules for probabilities.
>> Create, perform, and analyze results of probability simulations.
>> Create and interpret line plots, bar graphs, or histograms.
>> Calculate summary statistics for central tendency and variability.

What Students Will Learn

>> Apply knowledge of statistics and probability to solve challenging problems.

Problem #5

Ms. Butler has set up a contest for the 100 seventh-grade students at Morrison Junior High. Each day she chooses a whole number randomly from the set {1, 2, 3, 4, 5, 6} without telling anyone what it is. On the first day of the contest, every student in the seventh grade chooses a number between 1 and 6 and writes it with his or her name on a slip of paper that no one else sees. Students submit their slips. Those whose numbers match Ms. Butler's number are out of the game. Everyone else plays again the next day. The winner is the person who plays for the most days.

Directions

- Predict typical values of summary statistics for the number of days that each student plays: minimum, maximum, mean, median, mode, and interquartile range. Explain your thinking.
- Describe a simulation that would help you estimate these values. Talk about any assumptions that you make.
- Perform your simulation and graph the data. Describe the important features of your graph.
- Find the summary statistics for your data. Were your predictions on target? Explain.

CONVERSATION STARTERS FOR #5

What do you notice? What do you wonder?

Before performing the simulation:

I notice that the minimum is easy to predict and the maximum is hard to predict.
> It might make sense to predict a range of likely values for the maximum.

I notice that there is no limit to the number of days a student could play.

I notice that there could be some very high numbers and that this would increase the mean.

While performing the simulation:

I wonder what is the best way to record my data?
> Do you think it is important to list your results in any particular order? What is a practical way to record repeated values? (Consider tally marks.)

I notice that a substantial fraction of the students play for only a few days.

After performing the simulation:

I notice that it is easy to see the mode in the graph, but not as easy to see the mean or median.

I wonder what is the easiest way to calculate the mean from the table or graph?
> What would the list look like if you wrote every number?

I notice a connection between the median and the area under the graph.

SOLUTIONS FOR #5

Predictions for the minimum, maximum, mean, median, and mode: Students' answers will vary. The purpose of the question is to start them thinking about the situation.

A possible simulation: Roll a die as many times as you can until the number 1 (or another chosen number) comes up. Note that you can stick with the number 1; Ms. Butler had to pick a number randomly every day only so that no one could predict which number she would select. Write down the number of rolls, including the final roll in which you get the 1. Repeat this process many times. (Do it exactly 100 times if you want the simulation to look like the actual contest. Do it more times if you want to get a better estimate of the summary statistics.) This simulation assumes that students have a $\frac{1}{6}$ probability of predicting Ms. Butler's number each day, which will be true if she selects her numbers randomly.

Sample results for a simulation with 100 trials:

Number of Days Played	Frequency	Number of Days Played	Frequency	Number of Days Played	Frequency
1	21	11	2	21	1
2	16	12	1	22	0
3	11	13	1	23	0
4	10	14	1	24	2
5	6	15	1	25	2
6	7	16	0	26	0
7	3	17	2	27	1
8	4	18	2	28	1
9	0	19	2	29	0
10	2	20	1	30	0

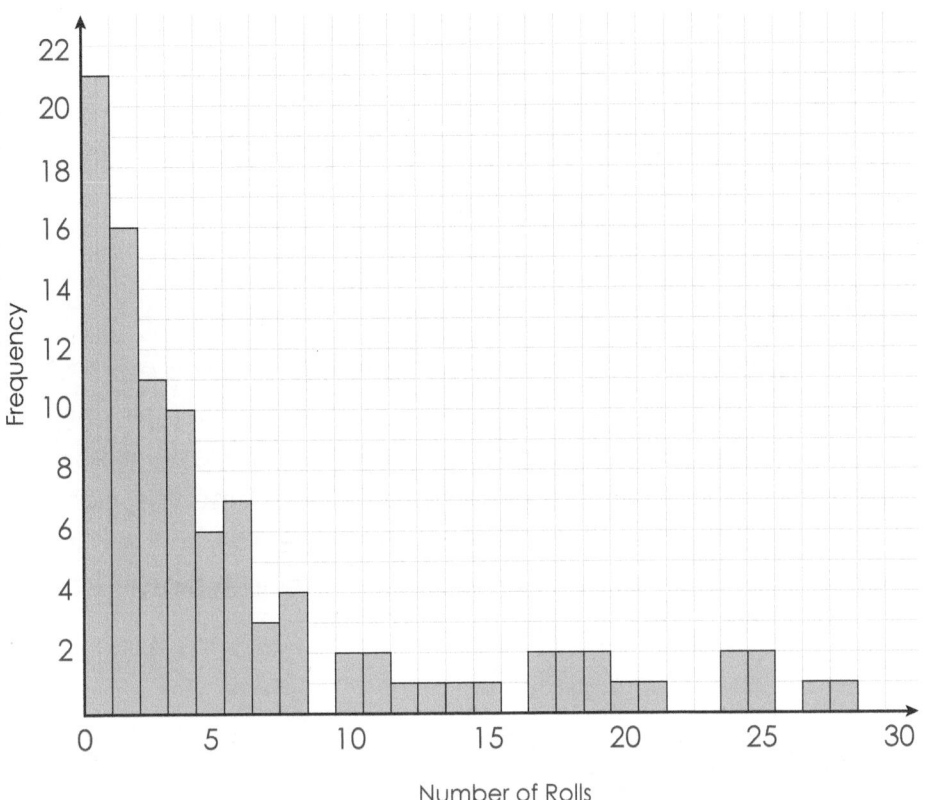

Number of Rolls

The frequency shows a clear tendency to decrease (more and more slowly) as the number of rolls (days) increases. The data cluster around numbers less than about 6. There are some gaps toward the right. Suppose that the data showed the results of an actual contest. Then, if there were more people playing the game, the graph would be likely to show a smoother, more consistent pattern of decrease. The gaps are probably due to the decreasing trend, and they would tend to fill in as the number of players increased. In this game, no one played for more than a month. However, there is no limit to how long a game could last! The graph could extend indefinitely far to the right.

Summary statistics: minimum: 1; maximum: 28; mean: 6.49; median: 4; mode: 1; interquartile range: 6 (between 2 and 8). Some students may be surprised that the mean, median, and mode are not all close to 6. In fact, these numbers are all close or equal to their theoretical values. (The mean is the only measure of central tendency that has a theoretical value of 6.)

Problem #6

Ms. Butler decides to set up the next contest so that about one fourth of the students earn a prize based on how long they can stay in the game.

Directions

- Calculate theoretical probabilities for playing each of 1, 2, 3, 4, 5, 6, 7, 8, 9, and 10 days. Explain your thinking.
- Compare your results to your data from Problem #5.
- Use both the theoretical probabilities and the data from Problem #5 to help Ms. Butler decide on the minimum number of days a student should typically stay in the game in order to win a prize.

Diving Deeper

Suppose Ms. Butler modifies the game so that each day, students earn points equal to the number of days they have been in the game so far. (For example, a student who has been in a game for 4 days has earned 1 + 2 + 3 + 4 = 10 points so far for that game.) But if their number ever matches Ms. Butler's, they lose all of the points they have earned for that game. At the end of each day, students decide whether to stay in the game or take their points and start a new game the next day. Students accumulate points all year long, and the one with the most points at the end of the year wins. Your job is to design strategies to maximize your chances of winning this game!

CONSERVATION STARTERS FOR #6

What do you notice? What do you wonder?

I notice that it is not hard to calculate the probability that a person plays for one day.

I notice that I need to build into my calculations the probability of *not* matching Ms. Butler's number.

I notice that the probability of playing for a certain number of days decreases more and more slowly as the number of days increases.

I notice that the probability can never reach 0 for any number of days.

I wonder what type of pattern in the probabilities can make this happen?

I wonder if it is possible for the probabilities to have a sum of 1 when there are an infinite number of them?

SOLUTIONS FOR #6

The theoretical probabilities for 1 through 10 days are approximately:

1: 16.7%	2: 13.9%	3: 11.6%	4: 9.6%	5: 8.0%
6: 6.7%	7: 5.6%	8: 4.7%	9: 3.9%	10: 3.2%

Thinking process: The probability of playing for one day is $\frac{1}{6}$, or about 16.7%. The probability for 2 days looks like $P(\text{"no match"} \text{ and } \text{"match"})$, because you must not match Ms. Butler's number on the first day. Using the multiplication rule, this probability is $\frac{5}{6} \cdot \frac{1}{6} = \frac{5}{36}$.

As you play for more days, you must continue not to match Ms. Butler's number until the final day. Therefore, the probability of playing for 3 days is $\frac{5}{6} \cdot \frac{5}{6} \cdot \frac{1}{6} = \frac{25}{216}$.

In general, the probability of playing for n days is $\left(\frac{5}{6}\right)^{n-1} \cdot \frac{1}{6} = \frac{5^{n-1}}{6^n}$. Notice that you may obtain each probability from the preceding one by multiplying it by $\frac{5}{6}$.

Comparison to the data: The fact that each successive probability becomes $\frac{5}{6}$ as great validates the observations of a mode of 1 and the decreasing frequency for greater numbers of days. Also, the probability will never reach 0, as expected. The simulated percentages for 1, 2, 3, 4, and 5 rolls (days) were 21%, 16%, 11%, 10%, and 6%, which are reasonably close to the calculated values. They would tend to become closer as you repeated the simulation more times.

Creating a rule to win about 25% of the time: Using the data, in 74 out of 100 games, a player remained in the game for 7 or fewer days. This suggests making a rule that a player wins by staying in the game for more than 7 days.

Using the calculated probabilities, you would expect to play for 7 days or fewer approximately 72.1% of the time, while you would play 8 days or fewer about 76.7% of the time. Based on these numbers, you might change the rule so that you win by playing for more than 8 days, because 76.7% is a little closer to 75% than 72.1% is.

ALGEBRA CONNECTIONS

In the exploration One More Time! and in Stage 2 of this exploration, students encounter events that are not mutually exclusive—events that have outcomes in common. In these cases, you must be careful when calculating $P(A\ or\ B)$, because you may otherwise count outcomes more than once. The relevant formula is:

$$P(A\ or\ B) = P(A) + P(B) - P(A \cap B),$$

as discussed in the Algebra Connections for the One More Time! activity.

In future courses, students will also learn a formula for handling events that are dependent, such as what occurs when marbles or cards are not replaced before selecting again:

$$P(A\ and\ B) = P(A) \cdot P(B|A),$$

where $P(B|A)$ represents the probability that event B will occur given that event A has already occurred. In this exploration, students handle the situation intuitively without using the formula when they adjust the probabilities on the second and third draws.

Problems #5 and #6 deal with *geometric distributions*, which relate to *geometric sequences* and *geometric series*, concepts that students will study in algebra. Geometric sequences are (typically infinite) ordered lists of numbers in which successive numbers have the same ratio. In Problem #6, the probabilities for 1 day, 2 days, 3 days, and so on, form a geometric sequence starting with $\frac{1}{6}$ and having a common ratio of $\frac{5}{6}$:

$$\frac{1}{6}, \frac{5}{36}, \frac{25}{216}, \frac{125}{1296}, \ldots$$

Sums of numbers in a geometric sequence form a *geometric series*. Students will learn a formula for finding such sums when the ratio is between 0 and 1:

$$S = \frac{a}{1-r},$$

where a is the first number in the series, and r is the common ratio. Substituting $a = \frac{1}{6}$ and $r = \frac{5}{6}$ in this formula, you obtain:

$$S = \frac{a}{1-r} = \frac{\frac{1}{6}}{1-\frac{5}{6}} = \frac{\frac{1}{6}}{\frac{1}{6}} = 1,$$

showing that the sum of the probabilities is 1 as expected!

Exploration **7**

Paths and Pascal

This exploration introduces students to the concepts of *permutations* and *combinations* by way of a classic problem that involves counting paths between two locations. These ideas are part of a branch of mathematics known as *combinatorics*, which is essentially about finding clever ways to count things. Although this might sound trivial at first, students who have experienced the challenge of counting outcomes for complex probability problems may appreciate the fact that counting is not always as easy as you expect!

The name "Pascal" in the title of the exploration refers to the 17th-century French mathematician, scientist, and philosopher, Blaise Pascal. Pascal made important contributions to the study of pressure and vacuum in physics, and was one of the first involved in the development of mechanical calculators. He also laid the groundwork for the development of the mathematical study of probability. Related to this effort was a famous triangle that bears his name. Students will discover this triangle and explore some of its fascinating patterns and practical uses in this activity.

DOI: 10.4324/9781003232780-10

STAGE 1

In Problem #1, students begin exploring a "number of routes" situation by counting paths to find probabilities. As they progress through the exploration, they will extend and apply what they learn from the problem by identifying and extending patterns. Some students will work just with pencil and paper. Others may prefer to act out the problem by flipping a coin.

As your students get engaged in the task, ask them to talk about how they are recording their work. Some may be drawing the paths, often trying to fit all of them into a single diagram, which makes it very hard to read! Encourage them to think of ways around this difficulty. They may draw a separate picture for each path. Or they may decide to represent each path with a "code," such as a string of letters that describes the direction of travel (e.g., EESSE) or the outcome of flipping the coin (HHTTH).

What You Will Need

 » Graph paper (for students who want to draw the paths)

What Students Should Know

 » Use lists, tables, and tree diagrams to record and organize outcomes.
 » Find probabilities for events with equally likely outcomes.

What Students Will Learn

 » Record and organize outcomes for complex probability problems.
 » Recognize and make use of patterns when solving problems.
 » Recognize and make use of symmetry when solving problems.

Problem #1

Home

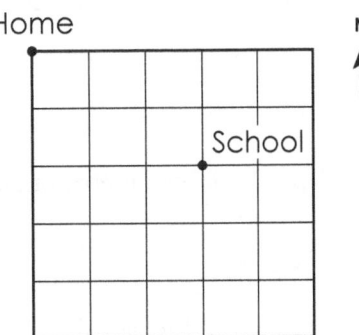

Start at home. Flip a coin 5 times. Each time you get "heads," travel one block east. Each time you get "tails," travel one block south.

Directions

- Show all of the places you could be at the end of this journey.
- Find the probability that you will end up at school.
- Explain your thinking.

Testing the Waters

Begin by answering the question for some of the intersections closer to home.

CONVERSATION STARTERS FOR #1

What do you notice? What do you wonder?

I notice that the paths never "backtrack."
> Every path is a shortest possible route along the streets to a given location.

I notice that the probability depends on the number of possible paths to school.

I notice that each path is equally likely.

I wonder if it would help to draw any of the paths?

I notice that I can apply what I have learned about using tree diagrams.

I wonder if I can predict the number of outcomes in the sample space without listing them?

I notice that the number of paths to school is equal to the number of paths to the intersection diagonally south and west of it.

I notice symmetry in the numbers of paths to certain intersections.

I wonder if the symmetry can lead me to an efficient strategy?

SOLUTIONS FOR #1

You could end up at any of the locations shown with a dot. Ask students to show the places you could land after 4 flips of the coin. What about after 6 flips? What is the pattern?

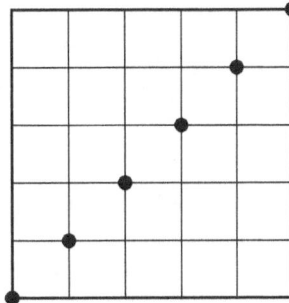

The probability of ending up at school is $\dfrac{10}{32} = \dfrac{5}{16}$.

Strategy 1: Make an organized list or a table of all of the outcomes. Each outcome represents a possible path. Of course, there are many ways to organize them. (Some students may also record H and T for "heads" and "tails" instead of E and S.)

EEEEE	ESEEE	SSSSS	SESSS
EEEES	**ESEES**	SSSSE	SESSE
EEESE	**ESESE**	SSSES	SESES
EEESS	ESESS	SSSEE	**SESEE**
EESEE	**ESSEE**	SSESS	SEESS
EESES	ESSES	SSESE	**SEESE**
EESSE	ESSSE	SSEES	**SEEES**
EESSS	ESSSS	**SSEEE**	SEEEE

In order to get to school, you must travel three blocks east and two blocks south (in any order). Therefore, the outcomes in this event contain three E's and two S's. There are 10 of these, and they are shown in bold. Because there are 10 outcomes in the event and 32 outcomes in the sample space, the probability is $\dfrac{10}{32}$.

Strategy 2: Begin with a tree diagram showing the first few blocks traveled.

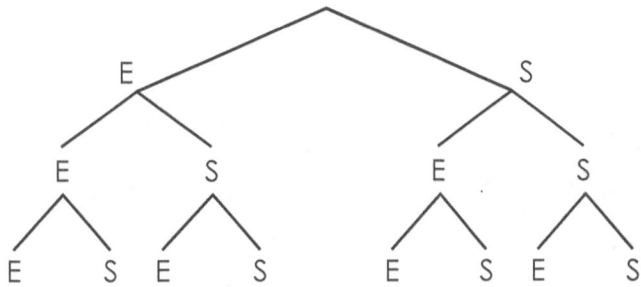

The number of outcomes doubles each time: 2, 4, 8, 16, 32. Therefore, the number of outcomes in the sample space is 32. With or without using a tree diagram as a guide, create an organized list of outcomes in the event. One method is to find every possible way to place two S's within a set of five blanks:

$$
\begin{array}{cccc}
\underline{S}\ \underline{S}\ \underline{\ }\ \underline{\ }\ \underline{\ } & \underline{S}\ \underline{\ }\ \underline{S}\ \underline{\ }\ \underline{\ } & \underline{S}\ \underline{\ }\ \underline{\ }\ \underline{S}\ \underline{\ } & \underline{S}\ \underline{\ }\ \underline{\ }\ \underline{\ }\ \underline{S} \\
\underline{\ }\ \underline{S}\ \underline{S}\ \underline{\ }\ \underline{\ } & \underline{\ }\ \underline{S}\ \underline{\ }\ \underline{S}\ \underline{\ } & \underline{\ }\ \underline{S}\ \underline{\ }\ \underline{\ }\ \underline{S} & \\
\underline{\ }\ \underline{\ }\ \underline{S}\ \underline{S}\ \underline{\ } & \underline{\ }\ \underline{\ }\ \underline{S}\ \underline{\ }\ \underline{S} & & \\
\underline{\ }\ \underline{\ }\ \underline{\ }\ \underline{S}\ \underline{S} & & &
\end{array}
$$

Fill the remaining slots with E's, and you have the 10 outcomes!

Strategy 3: The probability of traveling east five times is $\frac{1}{2} \cdot \frac{1}{2} \cdot \frac{1}{2} \cdot \frac{1}{2} \cdot \frac{1}{2} = \left(\frac{1}{2}\right)^5 = \frac{1}{32}$,

meaning that there are 32 outcomes in the sample space (and that there is only one way to travel east five times). To find the number of outcomes in the event, begin by counting the paths to get to other locations that are five blocks from home. There are five ways to travel east four times (EEEES, EEESE, EESEE, ESEEE, SEEEE). Similar calculations apply to traveling south five times and four times. Showing the number of paths on the map reveals a diagonal line of symmetry (the dotted line).

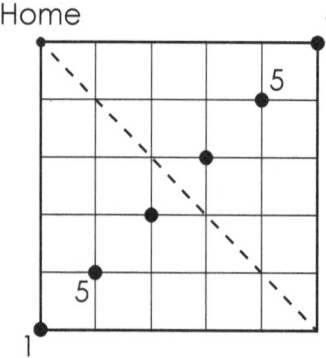

There are 32 paths associated with the large dots. 12 of them (1 + 5 + 5 + 1) have been accounted for. There are 32 − 12 = 20 of them remaining. Due to the symmetry, the remaining two locations (one of which is the school) must have 10 paths each.

STAGE 2

In Problem #2, students explore patterns from Problem #1 in order to discover the famous Pascal Triangle.

```
                1
             1     1
          1     2     1
       1     3     3     1
    1     4     6     4     1
 1     5     10    10    5     1
```

This is only a small piece of the actual triangle. Once you understand certain key patterns, you may easily add rows for as long as you like. Be aware that your students' work will look a little different than this picture, because their numbers will be arranged to form *right* triangles. However, their triangles will contain the same numbers as above.

The Pascal Triangle is named after the 17th-century mathematician Blaise Pascal, who discovered many uses for it (including in the area of probability). However, the numbers and patterns in the triangle were known and used in India, Persia, and China many centuries before Pascal lived. What we call the Pascal Triangle is known by many different names around the world. Mathematicians have found countless beautiful and useful patterns within it. Your students will discover its usefulness for counting outcomes in certain situations. And they will encounter it again in algebra when they study the *binomial theorem*.

Problem #3 in Stage 2 forms a bridge between Problems #2 and #4. Once your students have completed it, they will be prepared to take a new look at the Pascal Triangle and develop procedures—perhaps even formulas—for calculating any number in the triangle.

What Students Should Know

» Solve Problem #1 from this exploration.

What Students Will Learn

» Discover the Pascal Triangle and many patterns within it.
» Understand connections between the Pascal Triangle and probability.

Problem #2

Home

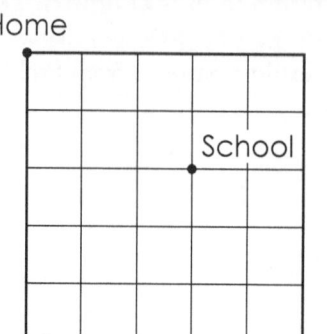

N

School

The probability of reaching an intersection (assuming that you flip the coin an appropriate number of times) depends on the number of east/south paths by which you can reach it from home. It is easier to count the paths for some intersections than for others.

Directions

- Find the number of paths to each intersection that is 5 blocks or fewer from home.
- Organize your answers. Look for patterns and describe them.
- Predict the number of paths to more distant intersections. Explain your thinking.
- Test some of your predictions.

Diving Deeper

Figure out how to create and extend the drawing below. How does it relate to Problem #2?

CONVERSATION STARTERS FOR #2

What do you notice? What do you wonder?

I notice an extremely simple pattern on the top and left edges of the grid.

I notice another simple pattern on the second row and second column.

I notice that the rows are identical to the corresponding columns.

I notice symmetry in the grid.

I wonder if the symmetry is the reason for the connection between rows and columns?

I notice a pattern when I add the numbers along the diagonals.

I notice a reason for this pattern related to tree diagrams for flipping coins.

I wonder if I can use my answers for 5-block intersections to find probabilities for 6-block intersections?

I notice that when I continue the grid, the numbers of paths get large very quickly!

I notice that I can predict the number of paths to an intersection by looking only at numbers in nearby intersections.

SOLUTIONS FOR #2

The number of paths for each intersection:

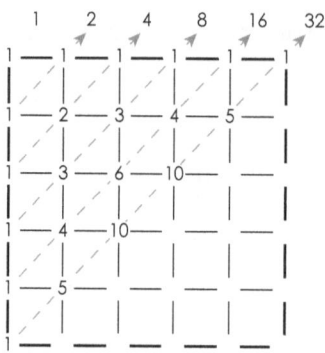

Some patterns:

>> The top row and left column are all 1s.

>> The second row from the top and the second column from the left are the counting numbers.

>> The third row from the top and the third column from the left are the *triangular* numbers. The differences between successive numbers are 2, 3, 4, 5, and so on.

>> The numbers are symmetrical across a line through the upper left and lower right corners of the map.

>> The sums along the diagonals are the powers of 2:

$$2^0 = 1 \quad 2^1 = 2 \quad 2^2 = 4 \quad 2^3 = 8 \quad 2^4 = 16 \quad 2^5 = 32 \text{, etc.}$$

>> Each number inside the grid is the sum of the number to the left of it and the number above it. (This pattern is your best key to making predictions!)

This triangular grid of numbers continues forever and is called the *Pascal Triangle*. See the Introduction to Stage 2 in order to learn more about it. *Note*: Students may discover many more patterns in the Pascal Triangle!

Sample predictions: The next diagonal of numbers (for locations that are 6 blocks from home) will read 1, 6, 15, 20, 15, 6, 1.

Testing the predictions: Students may test their predictions by listing paths and counting them or by checking that the sums along the diagonals are powers of two. For example, $1 + 6 + 15 + 20 + 15 + 6 + 1 = 64$, which is the next power of two: $2^6 = 64$.

A note about the Diving Deeper question: You may create this complex design simply by extending the grid and placing a circle on every even number!

Problem #3

Madison is collecting action figures and displaying them in a row on a bookshelf in her bedroom. So far she has collected four figures, and she is deciding how to arrange them.

Directions

- Find the number of different ways she can arrange the four action figures.
- Develop a method for finding the number of ways to arrange any number of action figures.
- Explain why your method works.

Diving Deeper

In the game Yahtzee, players roll five 6-sided dice. A "long straight" (which earns 40 points) contains five consecutive numbers. What is the probability of rolling a long straight in one roll? What about a "small straight," which contains four consecutive numbers?

Testing the Waters

Begin with fewer action figures.

CONVERSATION STARTERS FOR #3

What do you notice? What do you wonder?

I wonder if it would help to act the problem out with real objects?

I notice that it is very important to list the rearrangements in an organized way.

I notice that it helps to begin with a smaller number of action figures (as suggested in Testing the Waters).

I notice that it helps to begin with one action figure, increase the number one step at a time, and look for patterns.

I notice that knowing the number of arrangements for n figures helps me to find the number of arrangements for $n + 1$ figures.

I notice that, as I increase the number of action figures, the number of possible arrangements becomes very large very quickly!

SOLUTIONS FOR #3

Madison can arrange her 4 action figures in 24 ways.

Thinking process: One action figure may be arranged in only one way. Two action figures may be arranged in two ways: 1 2 or 2 1. Three figures can be arranged as follows:

1 2 3   1 3 2   2 1 3   2 3 1   3 1 2   3 2 1

Each pair of arrangements begins with a different number. There are $2 \cdot 3 = 6$ arrangements. For four figures, you have:

1 2 3 4	2 1 3 4	3 1 2 4	4 1 2 3
1 2 4 3	2 1 4 3	3 1 4 2	4 1 3 2
1 3 2 4	2 3 1 4	3 2 1 4	4 2 1 3
1 3 4 2	2 3 4 1	3 2 4 1	4 2 3 1
1 4 2 3	2 4 1 3	3 4 1 2	4 3 1 2
1 4 3 2	2 4 3 1	3 4 2 1	4 3 2 1

Each column of 6 begins with a different number, creating 4 groups of 6, which equals 24 arrangements. You could have predicted this without listing every arrangement! For each starting number, you know that there must be 6 ways to fill the remaining 3 spots, because you have already discovered that there are 6 ways to arrange 3 objects!

Patterns and predictions: Mathematicians call a "rearrangement" of objects a *permutation*. Based on what has happened so far, there must be 120 permutations of 5 objects, because for each of the 5 possible starting objects, the remaining 4 objects may be rearranged in 24 ways, and $24 \cdot 5 = 120$. Summarizing the results so far:

Number of Objects	Permutations	Calculation
1	1	1
2	2	$1 \cdot 2$
3	6	$2 \cdot 3$ (or $1 \cdot 2 \cdot 3$)
4	24	$6 \cdot 4$ (or $1 \cdot 2 \cdot 3 \cdot 4$)
5	120	$24 \cdot 5$ (or $1 \cdot 2 \cdot 3 \cdot 4 \cdot 5$)

To calculate the number of permutations of n objects, simply multiply all counting numbers from 1 to n. This is called "n factorial," and it is written with an exclamation point, "$n!$" In symbols, the definition of a *factorial* looks like

$n! = 1 \times 2 \times 3 \times \cdots \times (n-1) \times n$. "$n$ factorial" represents the number of ways in which n objects can be arranged. Factorials become extremely large very quickly. For example, try to find the value of 15!. The answer is greater than 1.3 trillion.

STAGE 3

Before beginning Problem #4, remind your students that the number of paths under discussion is actually the number of *shortest* paths (moving E and S only). Give them the Handout for Problem #4 before you give them the Problem page. Allow them plenty of time to look at the handout and think about it before you say anything. Solicit questions and observations, and ask if they can see any potential connections to the problems about the paths. After this discussion, you may hand out the Problem page in order to summarize and clarify the directions.

What You Will Need

» Copies of Handout for Problem #4 (in addition to the usual Problem Page)

What Students Should Know

» Understand the ideas from Stages 1 and 2 of this exploration.

What Students Will Learn

» Discover connections between factorials and the "paths" problem.
» Analyze complex patterns and use them to develop formulas.
» Understand why these formulas work.

Problem #4

The Handout for Problem #4 will help you develop a new strategy for counting the number of paths to any location on your grid.

Directions

- Find the number of occurrences of the string "EEE45."
- Find the number of occurrences of other strings on the right side of the handout. Describe what you notice, and explain why it happens.
- Use your observations to calculate the number of *different* strings on the right side of the handout. Explain your thinking.
- Imagine replacing all fours and fives with "S." Determine the number of different strings now. Explain your thinking.
- Invent and describe a quick method to predict the number of paths leading to any location in the grid.

Diving Deeper

Calculate the number of five-card hands than can be dealt from a 52-card deck. (Imagine laying out the 52 cards and labeling the cards with "Y" for *in the hand* or "N" for *not in the hand*.)

HANDOUT FOR PROBLEM #4

The left side of the page shows every permutation of the digits 1, 2, 3, 4, and 5. What has happened on the right side?

12345	21345	31245	41235	51234	EEE45	EEE45	EEE45	4EEE5	5EEE4
12354	21354	31254	41253	51243	EEE54	EEE54	EEE54	4EE5E	5EE4E
12435	21435	31425	41325	51324	EE4E5	EE4E5	EE4E5	4EEE5	5EEE4
12453	21453	31452	41352	51342	EE45E	EE45E	EE45E	4EE5E	5EE4E
12534	21534	31524	41523	51423	EE5E4	EE5E4	EE5E4	4E5EE	5E4EE
12543	21543	31542	41532	51432	EE54E	EE54E	EE54E	4E5EE	5E4EE

13245	23145	32145	42135	52134	EEE45	EEE45	EEE45	4EEE5	5EEE4
13254	23154	32154	42153	52143	EEE54	EEE54	EEE54	4EE5E	5EE4E
13425	23415	32415	42315	52314	EE4E5	EE4E5	EE4E5	4EEE5	5EEE4
13452	23451	32451	42351	52341	EE45E	EE45E	EE45E	4EE5E	5EE4E
13524	23514	32514	42513	52413	EE5E4	EE5E4	EE5E4	4E5EE	5E4EE
13542	23541	32541	42531	52431	EE54E	EE54E	EE54E	4E5EE	5E4EE

14235	24135	34125	43125	53124	E4EE5	E4EE5	E4EE5	4EEE5	5EEE4
14253	24153	34152	43152	53142	E4E5E	E4E5E	E4E5E	4EE5E	5EE4E
14325	24315	34215	43215	53214	E4EE5	E4EE5	E4EE5	4EEE5	5EEE4
14352	24351	34251	43251	53241	E4E5E	E4E5E	E4E5E	4EE5E	5EE4E
14523	24513	34512	43512	53412	E45EE	E45EE	E45EE	4E5EE	5E4EE
14532	24531	34521	43521	53421	E45EE	E45EE	E45EE	4E5EE	5E4EE

15234	25134	35124	45123	54123	E5EE4	E5EE4	E5EE4	45EEE	54EEE
15243	25143	35142	45132	54132	E5E4E	E5E4E	E5E4E	45EEE	54EEE
15324	25314	35214	45213	54213	E5EE4	E5EE4	E5EE4	45EEE	54EEE
15342	25341	35241	45231	54231	E5E4E	E5E4E	E5E4E	45EEE	54EEE
15423	25413	35412	45312	54312	E54EE	E54EE	E54EE	45EEE	54EEE
15432	25431	35421	45321	54321	E54EE	E54EE	E54EE	45EEE	54EEE

CONVERSATION STARTERS FOR #4

What do you notice? What do you wonder?

I notice that on the handout every 1, 2, and 3 has been replaced by E.

I notice that this makes the permutations look more like the descriptions of the paths.

I notice many patterns in the ways that the E's, 4s, and 5s are arranged.

I notice that every string on the right side of the handout appears the same number of times as every other string!

I wonder why this happens?

I notice that whenever the numbers 1, 2, and 3 occupy the same three positions in different strings (regardless of the order of the numbers), they become the same string when 1, 2, and 3 are replaced by E's.

I notice that I can divide to determine the number of different strings on the right side of the handout.

I notice that the right side of the handout shows every possible string with exactly three E's, one 4, and one 5.

After students have found a formula:

I notice that (because it counts paths), the formula can also be used to calculate the numbers in the Pascal Triangle!

I wonder what else this formula can be used for?

I wonder if there are formulas without factorial symbols that would work?

I wonder what 0! is equal to?

I wonder if I can use my formula to prove any of the patterns that I discovered in the Pascal Triangle?

SOLUTIONS FOR #4

The string "EEE45" occurs 6 times.

Occurrences of other strings: Every string on the right side appears exactly 6 times! This happens because the numbers 1, 2, and 3 were replaced by E's, and there are exactly 6 ways to arrange these three numbers.

The number of different strings: There are now 20 different strings, because by substituting 1, 2, and 3 with E, you replace the 120 different strings by 20 different groups of 6 identical strings each.

Replacing 4 and 5 with S: When you replace all of the fours and fives with S, you divide the number of different strings by 2, because every string that had the numbers 4 and 5 in the same position (even if they were in the opposite order) now looks identical. Because there were 20 different strings before, there are now 20 ÷ 2 = 10 different strings. This is consistent with the answer from Problem #1 for the number of paths to school!

A quick method for predicting the number of paths to any intersection: We began with the total number of permutations of 5 objects (because the path was 5 blocks long). The next step was to divide by 6 (the number of permutations of 3 objects), because replacing 3 different objects with E created groups of 6 identical strings. Finally, we divided by 2, because there are 2 ways to rearrange the 2 remaining objects that we replace by S. In short, we divided the factorial of the number of blocks in each path by the factorials of the number of E's and the number of S's.

If n is the number of blocks, and k is the number of E's, then the number of S's is $n - k$, and the formula for the number of paths is $P = n! \div k! \div (n - k)!$.

ALGEBRA CONNECTIONS

In second- or third-year algebra courses, students learn that the numbers in the Pascal Triangle are the *binomial coefficients*, which occur in the process of expanding algebraic expressions of the form $(x + y)^n$. They may also spend more time with these coefficients as they relate to *combinatorics* (counting problems). The expression:

$$n! \div k! \div (n - k)!$$

that appeared at the end of the Solutions for #4 is generally written:

$$\binom{n}{k} = \frac{n!}{k!(n-k)!}$$

and is used to count *combinations*: the number of ways to choose k objects from a set of n objects when the order in which the k objects appear is immaterial.

Suppose that we write the Pascal Triangle as shown in the introduction to Stage 2. If the top row of the Pascal Triangle is called the 0th row, and if the left-most number in each row is called the 0th element of that row, this formula also gives you the value of the kth element in the nth row of the Pascal Triangle. Try it! (Mathematicians have agreed to let $0! = 1$. Can you see why?)

An interesting challenge for advanced students who are comfortable with algebraic procedures is to prove the validity of the process for finding a number in the Pascal Triangle by adding two of the neighboring numbers. In symbols:

$$\binom{n}{k} = \binom{n-1}{k-1} + \binom{n-1}{k}$$

or equivalently,

$$\frac{n!}{k!(n-k)!} = \frac{(n-1)!}{(k-1)!(n-k)!} + \frac{(n-1)!}{k!(n-k-1)!}$$

Exploration **8**

Sports Correlations

In this exploration, students learn about using statistical concepts to analyze *bivariate* data—that is, data involving a relationship between two quantities, each of which may have variability.

The most common way to represent a bivariate relationship visually is to use a *scatterplot*. To create a scatterplot, you simply plot an ordered pair in a coordinate grid for each pair of numbers in the relationship. The Solutions in this exploration contain many examples of this process. Middle school math textbooks also provide details about the basics of creating and interpreting scatterplots. I am including a discussion in this introduction for those who find it helpful.

Often, when you have a pair of quantities that you are comparing, you think of one of them as depending on the other. The quantity on which one of them depends is known as the *independent* (or *control*) variable. It is shown on the horizontal axis and is traditionally represented by the letter *x*. The quantity that depends upon the control variable is called the *dependent* (or *response*) variable. It is shown on the vertical axis and is often represented by the letter *y*. In some cases, it will not be clear which variable belongs to which category, in which case you may choose.

When you interpret a scatterplot, you are looking for evidence of a connection, or *correlation*, between them. This will be reflected by patterns in the scatterplot. Because the data have variation, the patterns will not generally be precisely defined. You are simply looking for some predictability: dots that are closely aligned in some way or that show a trend toward rising or falling from left to right.

Correlations fall in a range from strong to weak and from positive to negative as illustrated below.

Strong Positive Correlation:
>> The graph shows a very clear upward trend from left to right.
>> When one variable increases, the other clearly tends to do the same.
>> Correlation coefficients: 0.7 to 1

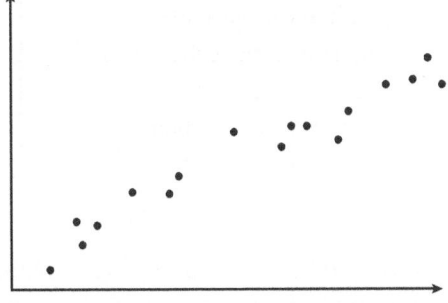

DOI: 10.4324/9781003232780-11

Moderate Positive Correlation:

> » The graph shows a notice-able upward trend from left to right.
> » When one variable increases, the other tends to do the same.
> » Correlation coefficients: 0.3 to 0.7

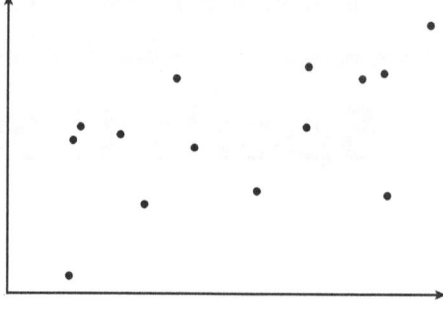

Weak Correlation:

> » It is hard to determine if the graph is rising or falling.
> » When one variable changes, little can be pre-dicted about the other.
> » Correlation coefficients: −0.3 to 0.3

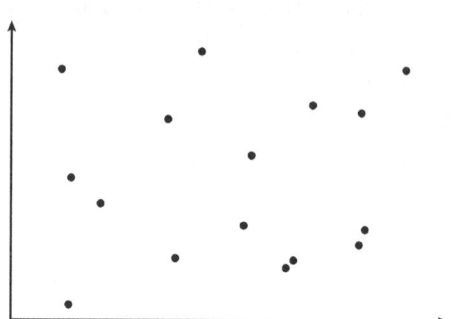

Moderate Negative Correlation:

> » The graph shows a notice-able downward trend from left to right.
> » When one variable increases, the other tends to decrease.
> » Correlation coefficients: −0.7 to −0.3

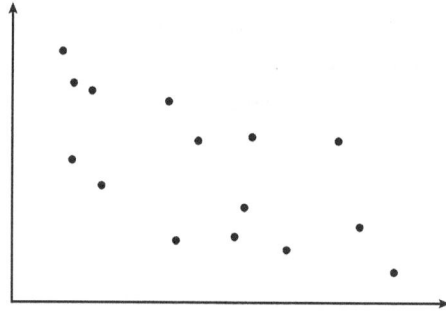

Strong Negative Correlation:

> » The graph shows a very clear downward trend from left to right.
> » When one variable increases, the other clearly tends to decrease.
> » Correlation coefficients: −1 to −0.7

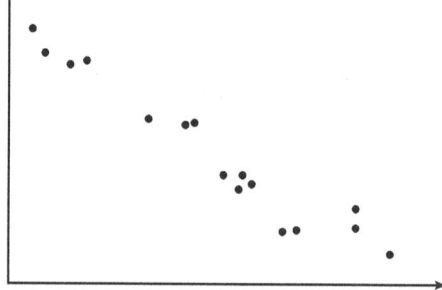

Information about *correlation coefficients* is included here for later reference. This concept is discussed in more detail in the Introduction to Stage 2.

STAGE 1

In Stage 1, your students will develop and apply their understanding of scatterplots and correlation to analyze statistics for the Women's National Basketball Association (WNBA) 2015 season. All of the data for the problems are shown on the next page. (*Note*: To enhance the learning experience, students may gather their own bivariate data for a different year, league, sport, or even another topic altogether. In this case, you can think of the teacher support features in this exploration as just a general guide to teaching the activity.)

Before beginning Problem #1, give your students time to familiarize themselves with the page of data and to discuss it as a class. What does each statistic mean? Find some pairs of variables that you believe will be strongly or weakly correlated and explain your reasoning. Will these correlations be positive or negative? Why? (Do not worry if this discussion "gives away" answers to some of the questions in Problem #1. Students will still be asked to write out their own explanations later.)

In Problem #2, your students are asked to create scatterplots for specific pairs of variables. Point out that they may have already completed some scatterplots in Problem #1 when they chose their own pairs.

What You Will Need

> » Women's National Basketball Association (WNBA) 2015 Team Statistics data page
> » Graph paper

What Students Should Know

> » Create scatterplots for bivariate data by plotting points in a coordinate grid.
> » Understand the concept of *correlation*. (See the Introduction to this exploration.)

What Students Will Learn

> » Increase skill and fluency with creating scatterplots.
> » Apply a general knowledge of correlation to interpret scatterplots.

Problem #1

The WNBA data on page 199 contain many variables with many relationships among them.

Directions

- Find a pair of variables that you think will have a positive correlation.
- Make a scatterplot for the variables.
- Decide if the scatterplot supports your prediction. Explain.
- Do the same for a negative correlation and for a correlation that is too weak to decide whether it is positive or negative.

Diving Deeper

Choose another league, year, or sport (or an interesting topic other than sports) that involves bivariate relationships. Gather data and explore the relationships in same way as in the directions above.

WOMEN'S NATIONAL BASKETBALL ASSOCIATION (WNBA)

2015 Team Statistics*

Team	W	FGM	FGA	3P%	FT%	Reb	Ast	Stl	Blk	TO	Pts	NR
Atlanta Dream	15	28.1	68.2	32.0	78.8	34.6	16.1	9.2	3.7	16.0	77.8	-2.8
Chicago Sky	21	31.5	70.5	32.7	82.6	36.6	15.9	6.7	6.7	12.1	82.9	5.2
Connecticut Sun	15	28.5	67.5	33.5	74.1	31.0	15.8	8.2	3.1	13.5	75.0	-2.2
Indiana Fever	20	27.6	65.1	36.0	79.9	32.4	14.9	8.7	2.9	14.4	77.7	2.4
Los Angeles Sparks	14	28.2	62.4	29.7	81.5	32.1	18.3	5.8	4.1	13.2	73.6	-1.0
Minnesota Lynx	22	28.9	65.7	33.2	81.7	35.3	18.1	6.2	4.4	13.1	75.5	5.4
New York Liberty	23	27.6	64.9	30.9	77.3	36.7	16.7	7.3	4.3	14.6	74.4	4.2
Phoenix Mercury	20	27.5	63.0	32.4	81.0	33.4	16.2	7.3	6.9	12.9	75.2	3.6
San Antonio Stars	8	25.3	64.9	29.5	77.7	32.6	15.2	7.2	3.8	14.0	68.1	-11.1
Seattle Storm	10	26.4	60.8	32.3	81.4	30.4	16.7	6.0	3.4	15.1	70.4	-7.5
Tulsa Shock	18	26.7	67.6	32.5	79.2	35.6	14.1	6.9	3.7	12.6	77.7	0.8
Washington Mystics	18	27.0	64.1	34.6	79.2	32.3	17.8	6.7	4.9	13.2	73.6	2.9

W: Wins

FGM: Field Goals Made per Game

FGA: Field Goals Attempted per Game

3P%: 3-point Percentage

FT%: Free Throw Percentage

Reb: Rebounds per Game

Ast: Assists per Game

Stl: Steals per Game

Blk: Blocks per Game

TO: Turnovers per Game

Pts: Points per Game

NR: Net Rating

*Note: Each team played 34 games in the 2015 season. Data retrieved from Women's National Basketball Association, 2016, http://ww.wnba.com/stats/team-stats.

CONVERSATION STARTERS FOR #1

What do you notice? What do you wonder?

I notice that most of the variables seem to be positively related to a team's success.

I notice that it may be relatively hard to find negative correlations.

I notice that the Turnovers variable is a likely candidate to be part of a negative correlation.

I notice that many of the correlations are weaker than I expected.

I notice that when I am trying to distinguish between positive and negative correlations, it often helps to imagine four regions in the coordinate grid: lower left, upper left, lower right, and upper right.

SOLUTIONS FOR #1

Pairs of variables with positive correlations: Examples of pairs that clearly have a positive correlation include W vs. NR, W vs. Pts, W vs. Reb, and FGM vs. FGA. However, there are numerous other possibilities.

Scatterplots: Scatterplots for these and other relationships are included among the Solutions to Problem #2. Students who choose pairs other than these may check each other's work. Alternatively, you may discuss some of them as a class.

Checking the scatterplots against predictions: Scatterplots of relationships with a positive correlation should show a general upward trend from left to right. In particular, there should generally be more points in the lower left and upper right portions of the scatterplot than in the upper left and lower right.

Pairs of variables with negative correlations: Examples of pairs that have a negative correlation may be more challenging to find. Consider looking at turnovers because they seem likely to have a negative impact on a team's success. Students may find other examples as well. Sometimes, the results may be surprising!

Checking the scatterplots against predictions: Scatterplots of relationships with negative correlations should show a general downward trend from left to right. In particular, there should generally be more points in the upper left and lower right portions of the scatterplot than in the upper right and lower left.

Pairs of variables with weak correlations: An example of a pair of variables with a weak correlation is W vs. FT%. Again, students may find other or better examples.

Checking the scatterplots against predictions: Scatterplots of relationships with weak correlations should show no clear upward or downward trend. In particular, points should appear roughly equally scattered between the upper left, upper right, lower left, and upper right regions of the scatterplot.

Note: Encourage students to extend their analysis beyond these basic points. There may be other interesting information hiding in their scatterplots! The Solutions to later problems give a few examples of this.

Problem #2

In some cases, you may be able to compare strengths of correlations visually. You may also be able to use scatterplots to make interesting observations that are not directly related to correlations.

Directions

- Make a scatterplot for each pair of variables:
  - Wins vs. Field Goal Percentage
  - Wins vs. Points
  - Blocks vs. Turnovers
  - Wins vs. Net Ratings
  - Wins vs. Rebounds
  - Wins vs. Turnovers
  - Points vs. Assists
  - Field Goals Made vs. Field Goals Attempted

- Based on the general appearances of the graphs, try to order them based on their correlations. Explain your ordering process and the thinking behind it.
- Choose two scatterplots that interest you. Study them and interpret what you see.

CONVERSATION STARTERS FOR #2

What do you notice? What do you wonder?

I notice that it is difficult to compare some correlations visually, because their strengths appear to be nearly the same.

I wonder which types of details I should include in my interpretation process?

I notice that one or two points may have a large effect on the appearance of a scatterplot and its correlation.

I wonder if I can calculate the field goal percentage directly from the field goals attempted and the field goals made *per game*?

> Yes.

I wonder if the increase in the number of turnovers (for most teams) *caused* the decrease in the number of games that they won?

> The two variables are *correlated* for those teams, but that does not necessarily mean that one thing *caused* the other. It certainly seems reasonable that having more turnovers will tend to decrease your number of wins, but it could also be that teams who have more turnovers struggle in other facets of the game as well—and these factors could also affect the teams' records.

I wonder why the W vs. NR scatterplot is not a perfect line?

SOLUTIONS FOR #2

Scatterplots:

Number of Wins vs. Field Goal Percentage

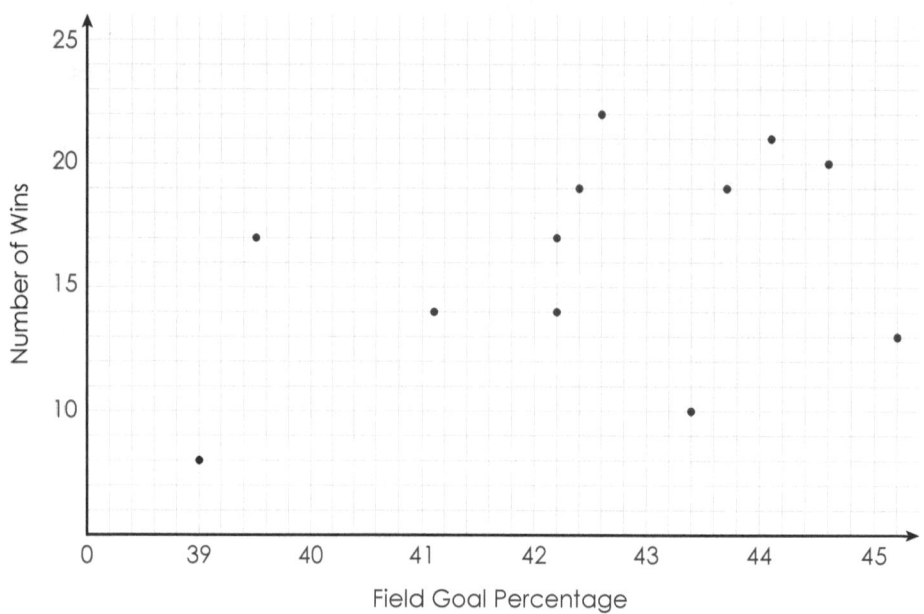

Number of Wins vs. Rebounds Per Game

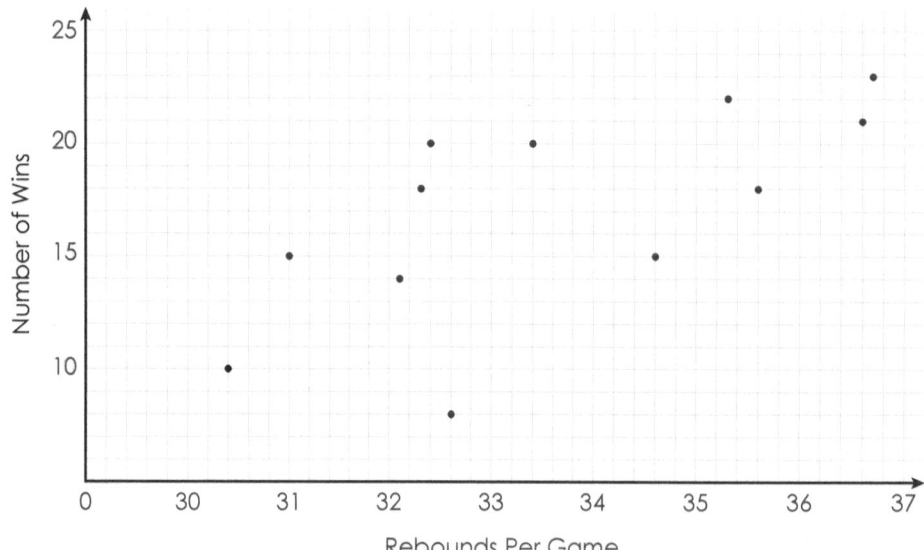

Number of Wins vs. Points

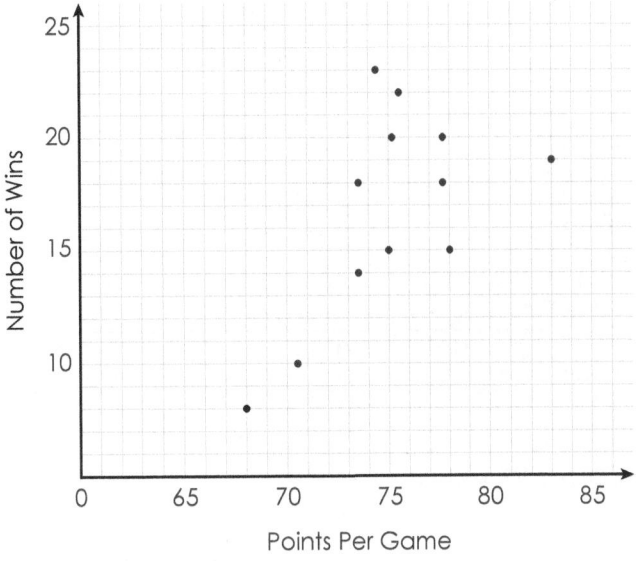

Number of Wins vs. Turnovers

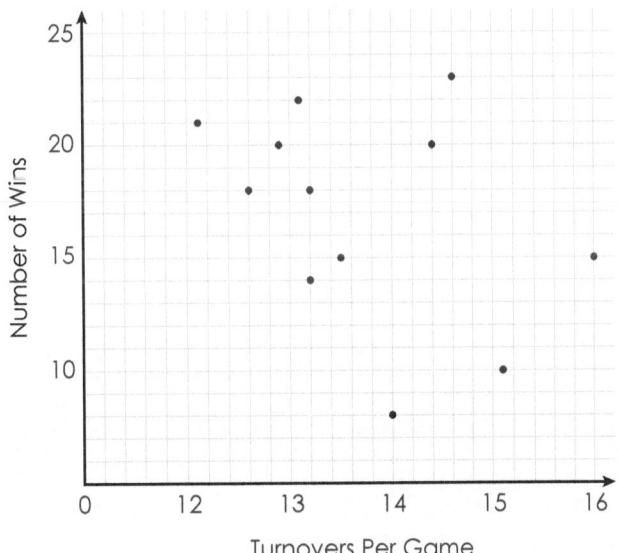

Blocks vs. Turnovers

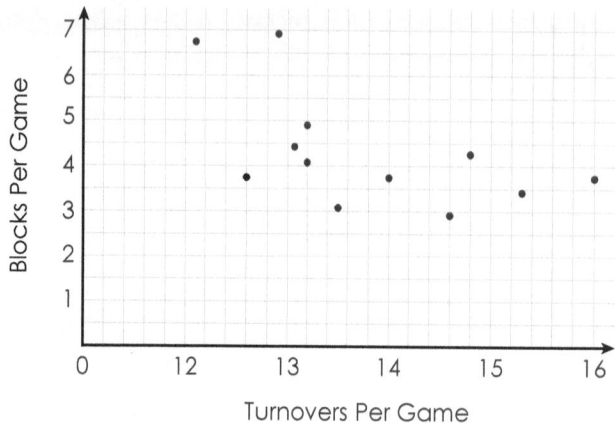

Points Per Game vs. Assists

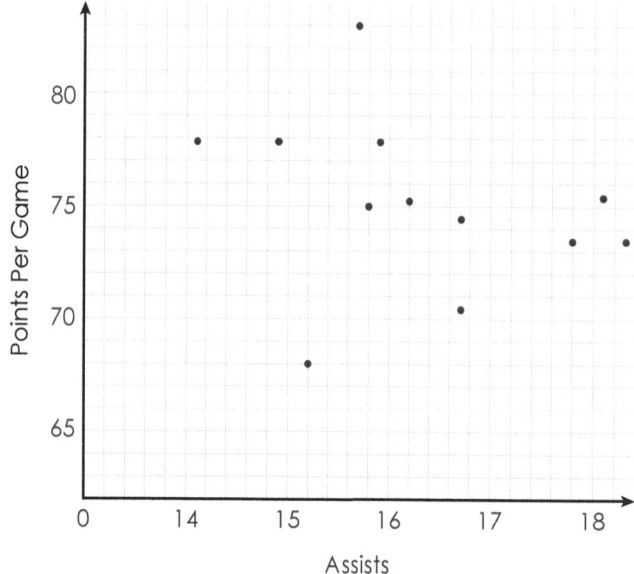

Number of Wins vs. Net Ratings

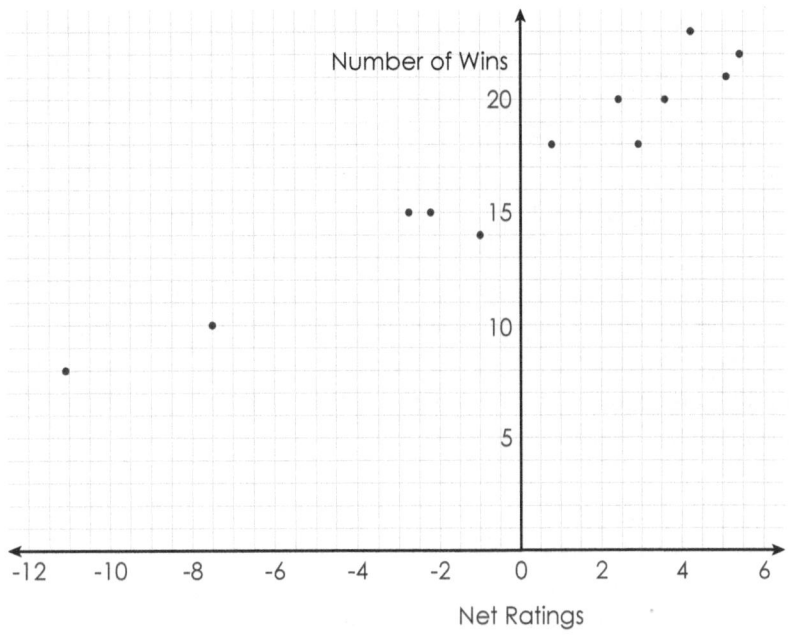

Field Goals Made vs. Field Goals Attempted

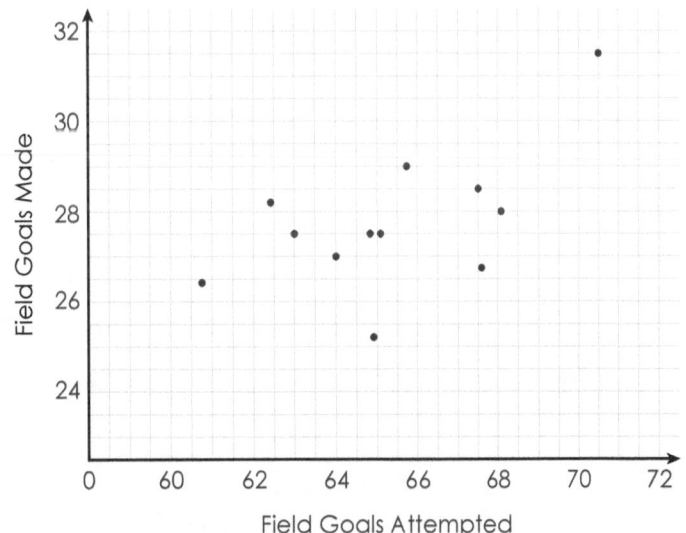

A possible ordering:

W vs. NR (very strong positive)

FGM vs. FGA (moderate to strong positive)

W vs. Pts (moderate positive)

W vs. Reb (weak positive)

W vs. FG% (weak positive)

W vs. TO (weak negative)

Bl vs. TO (weak to moderate negative)

P vs. Ast (moderate negative)

Students' answers will vary. The important idea is that scatterplots with stronger correlations have dots that are close together in such a way that it is easier to see an upward or downward trend. In upcoming problems, students will learn about methods used to calculate numbers (*correlation coefficients*) that quantify the strength and direction of the correlation. They will find that different methods may lead to a different ordering for the pairs! For example, one method shows W vs. Reb as a weak correlation while another shows it to be moderate to strong! See the introduction to Stage 2 of this exploration for more information about correlation coefficients.

In the sample above, the correlation coefficients run in the direction from strong positive to weak to strong negative. Some students may put them in the reverse order or arrange them in other ways. Regardless of how students organize their thinking, their answers should demonstrate their ability to distinguish positive vs. negative and weak vs. strong correlations.

Some sample interpretations:

Remember that students only need to choose two graphs to interpret. Their choices may not show up in the samples below. In any case, encourage them to discuss their interpretations in small groups and with the whole class.

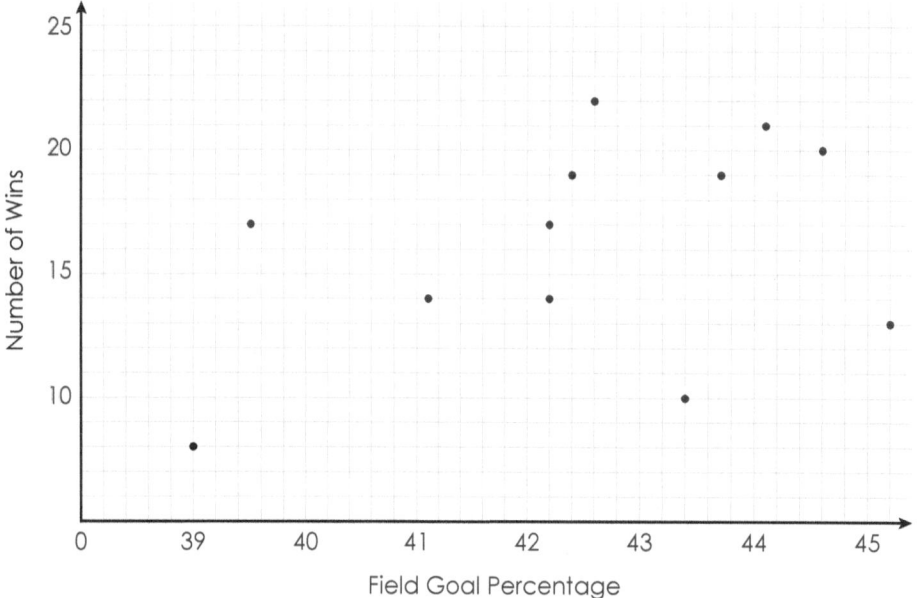

The Wins vs. Field Goal Percentage scatterplot appears to show a weak correlation. The points are quite spread out, and there are points in all regions of the graph (lower left, lower right, upper left, and upper right). The point in the far lower left and the relatively high concentration in the upper right may suggest a slight positive correlation, meaning that teams who make a higher percentage of their field goals may tend to win a few more games than other teams. However, it appears to be quite difficult to predict a team's Wins from its Field Goal Percentage.

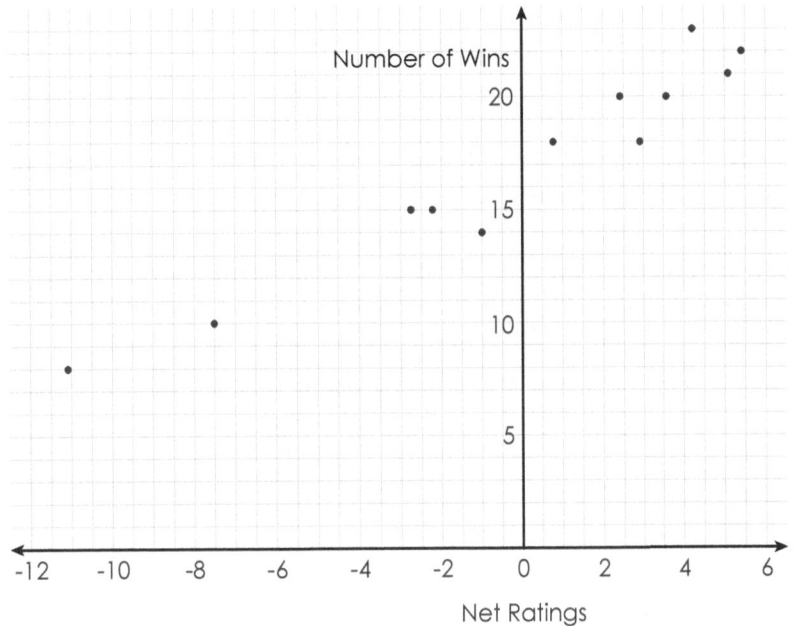

The Wins vs. Net Rating scatterplot shows a very strong positive correlation. The dots are close together following a clear upward trend, meaning that as the net rating increases, the number of wins increases at a relatively steady rate. The pattern is clear enough that the Wins can be predicted with reasonable confidence and accuracy from the Net Rating. It turns out that this makes sense, because the Net Rating takes both points gained and points allowed—the two quantities that directly determine the winner and loser of a game—into account. (See Problem #5 to learn more about Net Ratings.)

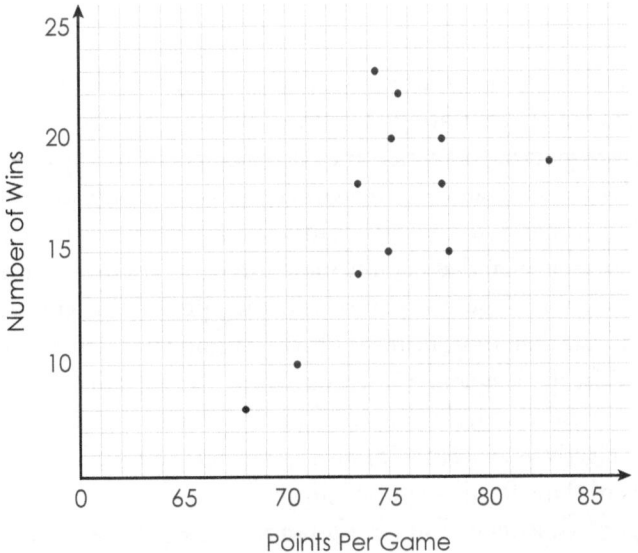

The Wins vs. Points scatterplot shows a fairly clear positive correlation (an upward trend), especially toward the left. Half of the points fall on a very clear line with a positive slope, showing that teams that score more points tend to win more games, as expected. However, the points that are farthest from the line show something interesting. Three of the points near the top center of the graph represent teams that won the most games but averaged only a moderate number of points per game. This provides evidence that the strongest teams relied heavily on defense to achieve their success. Going a step further, the graph appears to show that defense was consistently more important than offense when it came to winning at the highest level. None of the four highest scoring teams (the four points on the right of the graph) were in the top two winning teams (the two highest points on the graph).

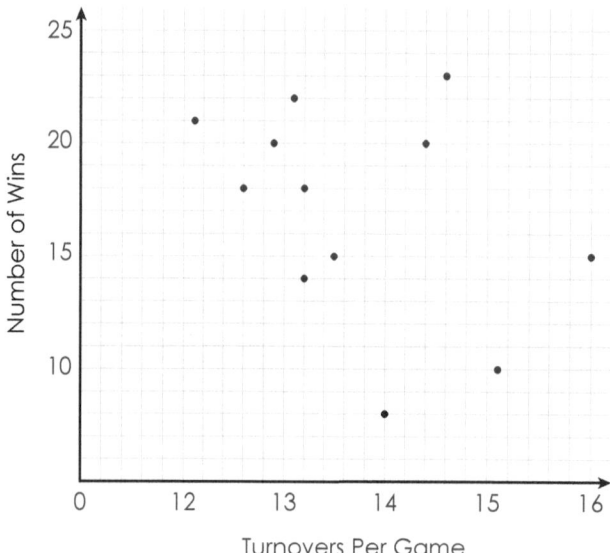

The Wins vs. Turnovers scatterplot appears to show a slight negative correlation (downward trend), indicating that teams who have more turnovers generally tend to win fewer games. This appearance depends partly on the lack of dots in the lower left region of the graph. On the whole, the correlation seems to be surprisingly weak, due primarily to the data points for three teams. The point farthest to the right represents a team that had the most turnovers yet managed to win a moderate number of games. The two points near the upper right represent teams that had more turnovers than most teams yet had very good win-loss records (including the best record in the league)! Apart from these three teams, the graph shows a very clear downward trend, showing that for the remaining teams, an increase in turnovers was clearly associated with having a worse record. In fact, the steepness of the downward slope for these points suggests that, for these teams, the decrease in the number of wins tended to be substantially greater than the increase in the number of turnovers (because one square represents the same change on both axes).

STAGE 2

In Stage 1, your students probably noticed how challenging it is to compare some correlations visually. Fortunately, mathematicians have developed methods for calculating numbers (*correlation coefficients*) that quantify the strength of a correlation. The *Pearson Correlation Coefficient*, represented by the letter r, is the most common of these. The formula for calculating r is complex, and the concepts behind it involve advanced algebra and calculus. However, students can use spreadsheet software, graphing calculators, or online tools to perform this computation.

The *Quadrant Count Ratio* (QCR) provides a simpler way to calculate correlation coefficients and usually gives results that are close to r. It focuses on four regions or *quadrants* in a scatterplot. (Some students may have discovered the importance of these regions in Stage 1!) To find the QCR, begin by calculating the mean for each variable. Draw a vertical line through the mean of the independent variable and a horizontal line through the mean of the dependent variable. This defines the four quadrants: Q1 (upper right), Q2 (upper left), Q3 (lower left), and Q4 (lower right).

Scatterplots with positive correlations tend to have more points in Q1 and Q3 than in Q2 and Q4. Those with negative correlations tend to have more points in Q2 and Q4. This leads to the following formula:

$$QCR = \frac{(\text{number of points in Q1 and Q3}) - (\text{number of points in Q2 and Q4})}{\text{Total number of points}}$$

Notice how the formula gives a positive QCR value for positive correlations, a negative value for negative correlations, and a QCR near 0 for weak correlations!

For example, the following scatterplot shows the relationship between blocks and turnovers. The mean number of turnovers is 13.725, and the mean number of blocks is 4.325. The lines through these numbers separate the grid into the four quadrants shown.

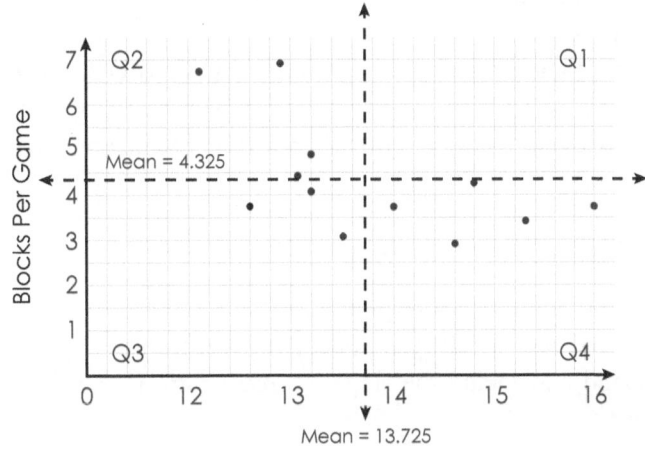

Turnovers Per Game

There are 0 points in Q1, 4 points in Q2, 3 points in Q3, 5 points in Q4, and 12 points in all. Therefore, the formula gives:

$$\text{QCR} = \frac{(0+3)-(4+5)}{12} = \frac{3-9}{12} = \frac{-6}{12} = -0.5$$

Notice how the formula captures the fact that there are 6 more points in Q2 and Q4 than in Q1 and Q3, resulting in the negative correlation.

The values of both the QCR and r are always between -1 and 1; -1 represents the strongest possible negative correlation, 0 represents no correlation, and 1 represents the strongest possible positive correlation. These relationships are summarized in the diagram below.

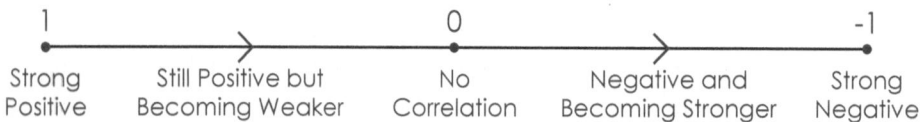

| 1 | Still Positive but | 0 | Negative and | -1 |
| Strong Positive | Becoming Weaker | No Correlation | Becoming Stronger | Strong Negative |

Your students should know that the process for calculating r is based on assuming that the relationship is roughly linear. Thus, the QCR may be a better measure of correlation than r in some situations when your scatterplot shows a pattern that is clearly not linear.

For scatterplots that do show an approximately linear relationship, the same spreadsheet or other tool that you use to find the value of r may also be used to find the *regression line* (or *line of best fit*). The regression line is the line that best approximates the relationship, and it is frequently used to make predictions. In Problems #4 and #5, your students will estimate, calculate, and interpret regression lines. If you do not have easy access to spreadsheets or other tools for calculating regression lines (or r), you may simply give your students this information. However, all students should estimate and interpret.

What You Will Need

» Graph paper
» Women's National Basketball Association (WNBA) 2015 Team Statistics data page
» Spreadsheet software or other tool for calculating regression lines (suggested)

What Students Should Know

» Create and interpret scatterplots.
» Understand and identify the slope, y-intercept, and x-intercept of a line.
» Know and understand the slope-intercept form ($y = mx + b$) of a linear equation.

What Students Will Learn

» Understand the meaning of a *correlation coefficient*.

» Calculate Quadrant Count Ratios.

» Use technology to find Pearson Correlation Coefficients and regression lines.

» Interpret correlation coefficients and regression lines, and use them to make predictions.

Problem #3

Knowing a formula for calculating strengths of correlations makes it easier to compare them.

Directions

- Calculate the Quadrant Count Ratio (QCR) for each of your scatterplots. Show at least one example of your calculation process.
- Use the QCR to order the graphs by their correlations, and compare the results to your estimated order from Problem #2. (If you see some differences, try to explain them!)

CONVERSATION STARTERS FOR PROBLEM #3

What do you notice? What do you wonder?

I notice that, because there are only 12 points in each scatterplot, there is a limited number of possible values for the QCRs.

In this case, each QCR will always be a decimal for a "12ths" fraction: approximately 0, 0.08, 0.17, 0.25, 0.33, 0.42, 0.50, 0.58, 0.67, 0.75, 0.83, 0.92, 1.00 or the opposites of these.

I notice that some of the QCR values contradict the order in which I placed my scatterplots in Problem #2.

This is to be expected. It will probably happen to most students, and it does not necessarily mean that your ordering in Problem #2 was incorrect, because there are different ways of measuring correlation.

I notice that some of the QCR values are close to or equal to each other.

Some correlations are so nearly equal in strength that even the use of correlation coefficients will not help you decide which is stronger. On the other hand, different methods for calculating correlation coefficients may distinguish them.

I wonder what other methods exist for calculating correlation coefficients?

I wonder if different methods for calculating correlation coefficients will always agree on which correlations are stronger?

SOLUTIONS FOR #3

The Quadrant Count Ratios (QCRs):

Variables	QCR
W vs. NR	$\frac{12-0}{12} = 1.00$
FGM vs. FGA	$\frac{10-2}{12} \approx 0.67$
W vs. Pts	$\frac{9-3}{12} = 0.50$
W vs. Reb	$\frac{8-4}{12} \approx 0.33$

Variables	QCR
W vs. FG%	$\frac{7-5}{12} \approx 0.17$
W vs. TO	$\frac{4-8}{12} \approx -0.33$
Pts vs. Ast	$\frac{3-9}{12} = -0.5$
Bl vs. TO	$\frac{3-9}{12} = -0.5$

Note: The QCRs are listed in order from 1 to –1. This agrees with the order shown for Problem #2. Other methods of calculating correlation coefficients will change this order in some cases. For example, the Pearson correlation coefficients (see the next problem) show the Bl vs. TO correlation to be quite a bit stronger than the Pts vs. Ast correlation even though they have the same QCR.

Sample calculation process:

Number of Wins vs. Field Goal Percentage

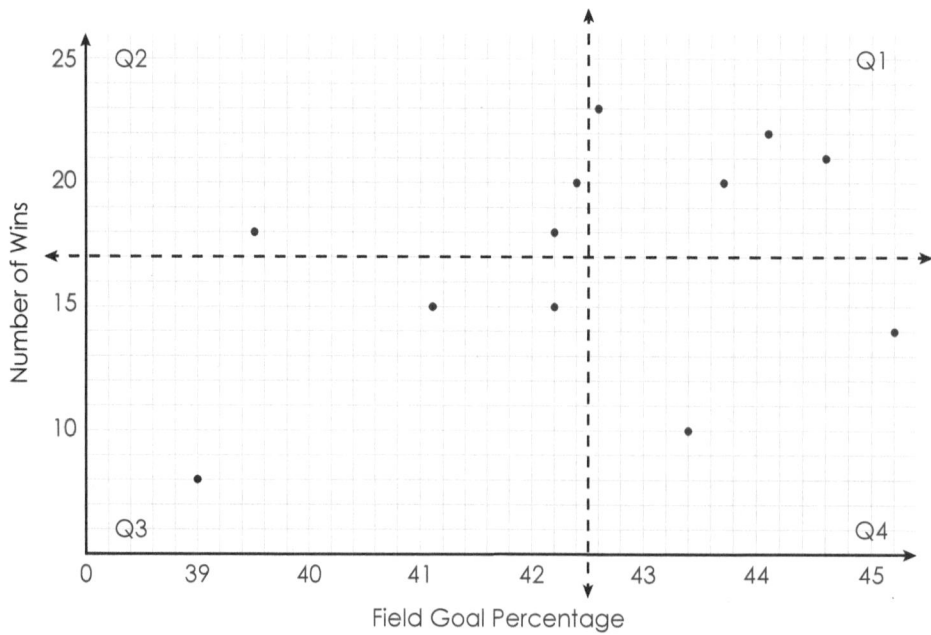

$$QCR = \frac{7-5}{12} \approx 0.17$$

There is a vertical dotted line at the mean field goal percentage, 42.5, and a horizontal dotted line at the mean number of wins, 17. These two lines divide the grid into the four quadrants Q1, Q2, Q3, and Q4.

There are 7 points in Q1 and Q3 and 5 points in Q2 and Q4. Thus, the difference between the number of points in Q1 and Q3 and the number of points in Q2 and Q4 is $7 - 5 = 2$. Because there are 12 points in all, the ratio of this difference to the total number of points is $\frac{2}{12} = \frac{1}{6} \approx 0.17$.

Problem #4

When the points in a scatterplot show a rough linear trend, you can draw a line that approximates the trend, and you can measure the correlation by determining how far the points in the scatterplot are from this line.

Directions

- Choose the scatterplot with the strongest correlation. Visually estimate and sketch the graph of a *regression line* (line of best fit) for it.
- Use your graph to estimate an equation for the regression line. Explain your thinking.
- Use the *linear regression* feature of a spreadsheet application, graphing calculator, or online tool to calculate an equation for the regression line. Compare it to your estimated equation.
- Use the same tool to identify the Pearson Correlation Coefficient (r), and compare it to the QCR for this relationship. (See Problem #3.)

Diving Deeper

Estimate and calculate lines of regression for some of the other scatterplots if you think they are close enough to being linear. Find their Pearson Correlation Coefficients. Comment on the values of the coefficients as they compare to the appearances of the graphs.

CONVERSATION STARTERS FOR #4

What do you notice? What do you wonder?

I wonder if the points would have to form a perfect line in order for r to equal 1?

> Yes. r does more than measure correlation. It determines how close the points are to the regression line. Thus, if there are any points that are not on the line, r cannot be exactly equal to 1.

I notice that it is possible to draw a line whose slope appears to show the correct trend of the correlation and that has an equal number of points above and below it.

I wonder if there is always an equal number of points above and below the regression line?

> Not necessarily, although it should usually be close. It also depends on how far the points are from the regression line.

I notice that I have to pay close attention to the scales on the x- and y-axes when I try to determine the slope of my estimated regression line.

I notice that the y-intercept of the regression line occurs at 17 games, which is half the number of games played in one season!

I wonder if this is a coincidence. If not, *I wonder* what the reason is?

I notice that whenever there are some points in Q1 and Q3, but none in Q2 or Q4, the QCR will always equal 1.

I notice that the scatterplot has a QCR of 1 even though the correlation is not perfect. (The points do not line up in a perfect pattern.)

I notice that r is close to but not equal to the QCR.

I wonder what the Net Rating statistic is and why it does such a good job of predicting the Number of Wins?

> See the Conversation Starters from Problem #5 to learn more about how a team's Net Rating is calculated and what it means.

SOLUTIONS FOR #4

The regression line for Wins vs. Net Rating:

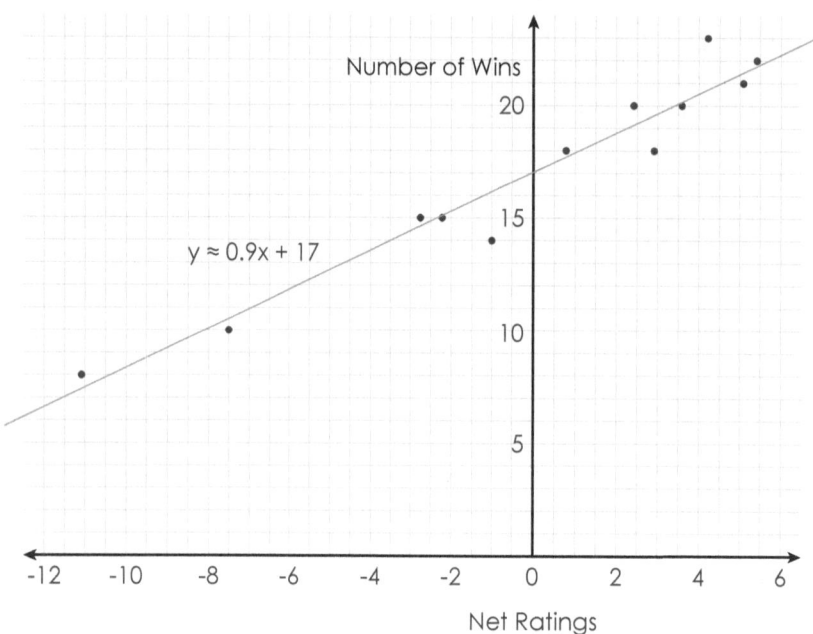

Although the estimated lines that your students sketch are unlikely to be exactly the same as the regression line, they should be reasonably close. Notice that in this case, there is a roughly equal number of points above and below the line. This does not always happen (depending on the distances of individual points from the line), but it is an interesting thing to watch for.

An equation for the line of best fit: An approximate equation for the line is $y = 0.9x + 17$. The *y*-intercept of 17 is fairly easy to see from the graph. Students may have different methods for estimating the slope of 0.9. One approach is to notice that the graph rises by approximately 0.9 units for every 1-unit shift to the right.

The Pearson Correlation Coefficient (r): The Pearson Correlation Coefficient is $r \approx 0.97$, which is very close to the value of 1.00 for the QCR. Both numbers indicate an extremely strong positive correlation. The QCR is exactly equal to 1, because all of the points are in Q1 and Q3. On the other hand, *r* is not quite equal to 1, because the scatterplot is not a perfect line.

Problem #5

The purpose of calculating regression lines and correlation coefficients is to answer questions and make predictions about real-world data.

Directions

- Interpret the line of best fit from Problem #4.
- Use the line of best fit to predict the number of wins for teams with other net ratings.
- Think about whether the approximate linear relationship is likely to continue for extreme values of the Net Ratings statistic. Explain your thinking.

CONVERSATION STARTERS FOR #5

What do you notice? What do you wonder?

I wonder what the Net Rating statistic means?

It is important to understand this meaning in order to interpret it! You calculate a team's Net Rating by subtracting its *Defensive Rating* from its *Offensive Rating*. The Offensive Rating is the number of points scored per 100 possessions, and the Defensive Rating is the number of points allowed per 100 possessions.

I wonder what is involved in interpreting lines of best fit and correlation coefficients?

The ultimate goal of interpretation is to figure out what the regression line and correlation coefficients are telling you about the real-world situation. The slope and y-intercept of the line are especially important to consider.

I wonder if a team could get a very high Net Rating without winning a lot of games?

Yes. A team could get a high Net Rating by winning games by many points and losing games by just a few points.

I wonder what positive and negative Net Ratings mean?

Try comparing each team's Net Rating to its win-loss record. You should notice something interesting (but not surprising).

I notice that I can use either the regression line's equation or its graph to make predictions.

It is a great idea to do it both ways in order to deepen your understanding and to check your results.

I notice that the regression line cannot continue forever, because no team can win less than 0 or more than 34 games.

SOLUTIONS FOR #5

Interpretation of the line of best fit: The equation $y = 0.9x + 17$ represents a line with a slope of 0.9 and a y-intercept of 17.

The positive slope indicates a positive correlation between the Number of Wins and the Net Rating. More specifically, the slope of 0.9 shows that each increase of 1 on the Net Rating scale corresponds to winning approximately 0.9 more games.

The y-intercept of 17 shows that teams with a Net Rating of 0 would be expected to win about 17 games, which is 50% of their games. This suggests a reason for using negative numbers in the Net Ratings scale. It appears that positive values correspond to winning teams and negative values correspond to losing teams.

Other predictions: It is interesting to think about the x-intercept as well. When $y = 0$, $0.9x + 17 = 0$. Solving for x gives a value of approximately −19 for the x-intercept, which means that a team with a Net Rating of about −19 would not be expected to win any games. However, this prediction may be a little questionable, because $x = -19$ is far away from the pattern in the scatterplot.

Choosing a point such as $x = -5$ may give a more dependable prediction. Using the graph, find −5 on the x-axis and go vertically upward to the line. From here, travel horizontally to the right until you reach the y-axis. The value at this point appears to be between 12 and 13.

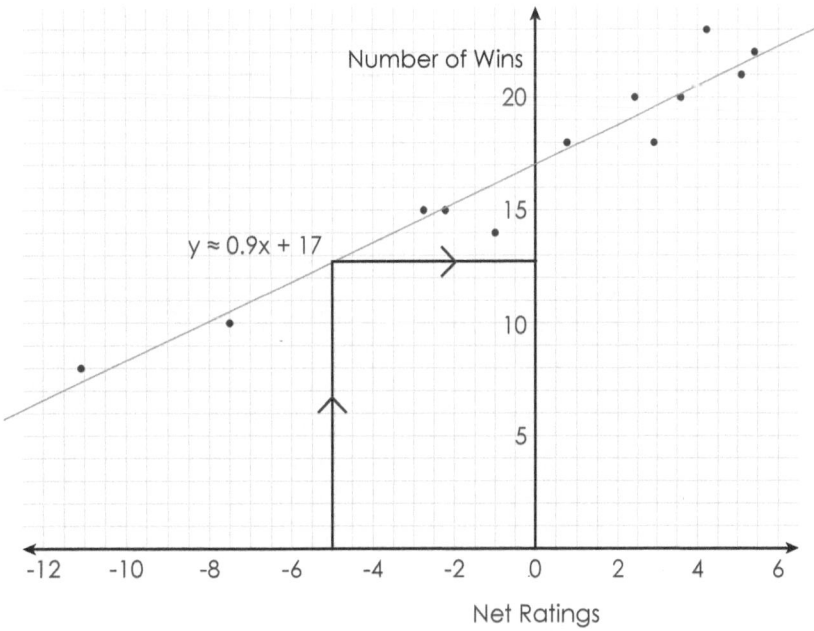

Therefore, we would predict with some confidence that a team with a Net Rating of −5 would be expected to win about 12 or 13 games.

223

Of course, you may also use the formula to make this prediction. Simply substitute −5 for x, and calculate:

$$y = 0.9x + 17$$
$$y = 0.9(-5) + 17$$
$$y = -4.5 + 17$$
$$y = 12.5$$

This is consistent with the answer obtained from the graph. (However, if you look carefully, you will see that the point on the y-axis appears to be slightly greater than 12.5. Why does this happen?)

Predictions for extreme values: It is tempting to make predictions about teams with very high (or very low) Net Ratings. For example, it is theoretically possible (although *extremely* unlikely!) that a team could have a Net Rating of 25. The formula predicts that this team will win $0.9(25) + 17 = 22.5 + 17 = 39.5$ games! Of course, this is impossible, because there are only 34 games in a season. The formula makes sense only for y values between 0 and 34. It seems reasonable to expect that near the extremes of this range, the accuracy of the formula's prediction may decrease. In other words, the relationship may not be as close to linear.

STAGE 3

In Problem #6, your students will create and test their own *metrics* to predict a team's overall "strength." You can think of a metric as a way of measuring something; it can be either a process or a formula. In advanced mathematics, a metric is often defined as something that measures distance. However, in real-world situations, you can attempt to design metrics that measure anything you like. The only catch is that they have to work!

The Net Rating statistic that your students explored in earlier problems is an example of a metric. In this problem, students will design their metrics by choosing statistics that they think are important and combining them into a single formula. Then, they will create and analyze scatterplots in order to test their metrics.

What You Will Need

- » Graph paper
- » Women's National Basketball Association (WNBA) 2015 Team Statistics data page
- » A tool for calculating regression lines and correlation coefficients (optional)

What Students Should Know

- » Understand the concepts from Stages 1 and 2.

What Students Will Learn

- » Apply their knowledge of bivariate statistics to create and test an original measurement tool.

Problem #6

The Net Ratings statistic is a *metric* that combines multiple statistics into a single equation in order to create a new statistic that measures a team's overall strength. The strong positive correlation with a team's Number of Wins provides evidence that the metric works very well.

Directions

- Create your own metric to predict a team's overall strength. Explain your thinking.
- Use your metric to calculate a value for each team.
- Evaluate the effectiveness of your metric. Justify your conclusions.
- Think of more questions to ask about your metric or other metrics.

Diving Deeper

Continue creating and exploring metrics. Compare them. Can you create one that appears to do a better job than the "Net Rating" metric of predicting the WNBA teams' number of wins? Are you confident that your metric would continue to work well in future seasons? Why or why not?

CONVERSATION STARTERS FOR #6

What do you notice? What do you wonder?

I wonder if I can use as many or as few statistics as I like?

Yes, although you should probably use more than one! The ideal metric might be one that is effective without being excessively complicated.

I wonder how to decide which statistics to include in my metric?

There are no rules for this. The data sheet provides many options. Choose statistics that you think may have the greatest relationship to a team's success. If you are curious about the importance of a statistic in predicting a team's strength, try including it to see how well it works!

I wonder if there are statistics not included on the data sheet that might be helpful.

Yes. In fact, statisticians have developed advanced statistics that involve adjusting simpler statistics in small ways to make them more meaningful. (Visit http://www.wnba.com to search for this data.) However, try not to use advanced statistics that already incorporate multiple statistics!

I wonder which operations I should use to combine my statistics?

Again, there are no clear-cut rules. However, you may discover that it makes sense to think of your equation in "chunks" in which each chunk represents a different statistic and the chunks are added or subtracted. Within each chunk, you might use multiplication or division to adjust how strongly that statistic contributes to the final value of your metric.

I wonder how good is "good enough" when testing my metric?

Once again, there is no rule that says what a good correlation coefficient would have to be. Obviously, the closer to 1 the better, but you have to be realistic. Finding a metric that works better than the Net Rating may be tough, because it uses points scored and allowed, which relate directly to winning and losing. On the other hand, you would probably hope that your metric does a better job than most individual statistics do alone.

I wonder if my metric will continue to work well for other leagues, years, etc.?

SOLUTIONS FOR #6

An example of a metric:

For the sake of illustration, I am sharing a metric that I created. Your students may explore this metric if they like, but they will learn more and have more fun if they create their own. They may even enjoy competing as individuals or teams to see who can create the most effective one!

My metric includes three parts: a typical shooting percentage (without free throws), rebounds, and turnovers. Just for fun, I named it the "Power Rating" (P). The formula is:

$$P = \frac{FG\% + 3P\%}{2} + Reb - 2 \cdot TO$$

The fraction on the left calculates a sort of overall shooting percentage by averaging the field goal and 3-point percentages. I tried to keep the typical values in each of the three parts of my equation roughly the same size so that each component carried approximately equal weight. (This was my reason for combing the shooting percentages and for doubling the turnovers.) I used my formula to calculate a Power Rating for each WNBA team.

Power Ratings (P) and Wins (W) for WNBA Teams in 2015

Team	P	W
Atlanta Dream	39.2	15
Chicago Sky	51.1	21
Connecticut Sun	41.9	15
Indiana Fever	42.8	20
Los Angeles Sparks	43.2	14
Minnesota Lynx	47.8	22
New York Liberty	44.3	23
Phoenix Mercury	45.7	20
San Antonio Stars	38.9	8
Seattle Storm	38.1	10
Tulsa Shock	46.4	18
Washington Mystics	44.3	18

Evaluating the effectiveness of the metric: To assess the effectiveness of my Power Rating, I investigated its correlation with the Number of Wins. I began with a scatterplot.

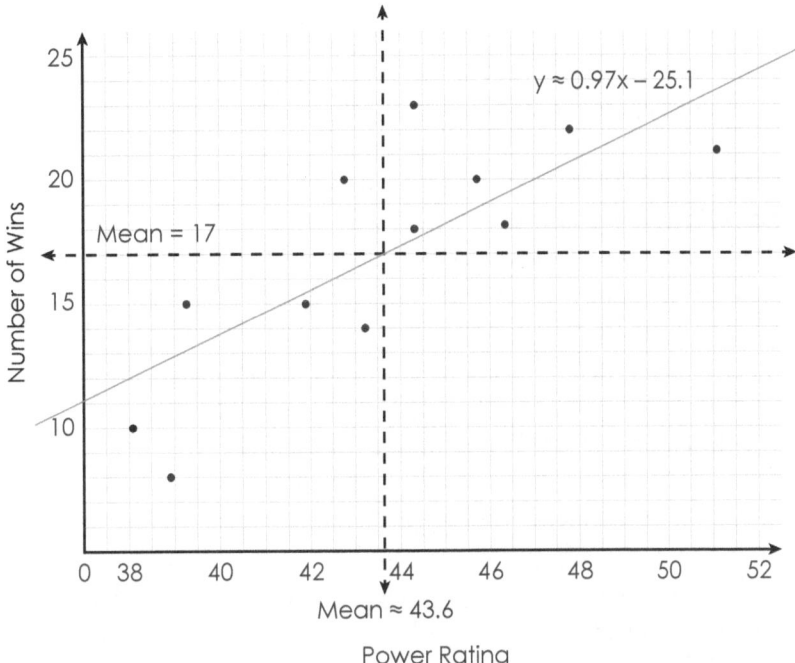

The general appearance of the scatterplot seemed to indicate a reasonably clear positive correlation. This was supported by a calculation of the QCR:

$$\frac{11-1}{12} = \frac{5}{6} \approx 0.83$$

The value 0.83 clearly represents a positive correlation. In fact, it is stronger than any of the correlations in the earlier scatterplots except for the Wins vs. Net Rating, which makes sense, because it includes more variables, some of which by themselves had moderately strong correlations with the Number of Wins.

I also used a spreadsheet to find the equation of the line of best fit ($y = 0.97 - 25.1$) and the Pearson Correlation Coefficient ($r \approx 0.79$), which provided further evidence for a reasonably strong positive correlation. Incidentally, some students may notice that the line of best fit appears to pass directly through the point where the two "mean lines" intersect! This seems reasonable, and, in fact, it is not a coincidence. Students will learn more about the reasons for this in future statistics courses.

I was interested to notice that the two points farthest from the line of best fit belong to the New York Liberty and the San Antonio Stars—the most and least successful teams in the league, respectively. It would be interesting to follow up on this observation with data from other years in order to see if something similar occurs.

In summary, the strong positive correlations provide evidence that my metric did quite a good job of predicting a team's Number of Wins. However, it did not seem to work as well for the strongest and weakest teams. In order to feel more confident that the Power Rating is a useful metric, it would be a good idea to test it for more seasons, past and future.

More questions to ask: There are virtually limitless possibilities for further questions to ask and explore. For example:

- » How much would the correlation coefficients change if the two "extreme" teams were removed from the calculations?
- » Would my metric be more or less effective if I did not multiply the number of turnovers by 2? What if I simply added the shooting percentages instead of finding their mean?
- » Could I improve my Power Rating formula (without changing the variables) by adjusting the strength of the contribution of each of the three parts—for example, by multiplying each part by a number greater than 1 to increase its contribution or a number less than 1 to decrease it?
- » Could I improve the Power Rating formula by incorporating more or different variables?
- » Can I find a metric that does a better job than the Power Rating? Can I find one that does a better job than the Net Ratings metric?
- » Will the Power Rating show strong correlations with the Number of Wins for other years?
- » Could I apply this Power Rating successfully to other basketball leagues, professional or otherwise?
- » Can I learn anything by exploring metrics that are not particularly successful in predicting a team's success? Why or why not?
- » Could I use different metrics to draw conclusions about the relative importance of different statistics in contributing to a team's success?

ALGEBRA CONNECTIONS

This exploration assumes a greater knowledge of elementary algebra than others earlier in the book. Students are already assumed to know about the meaning of the slope and y-intercept of a line and to be able to find an equation of a line using its slope-intercept form. However, it is possible for them to explore further by looking at some formulas.

I did not share equations for calculating the Pearson Correlation Coefficient (r) or lines of regression, because they are very complex, and students are not in a position yet to understand them conceptually. However, in case you have (courageous!) students who would enjoy playing around with it, I show a formula for r below:

$$r = \frac{N\sum xy - \sum x \sum y}{\sqrt{\left[N\sum x^2 - \left(\sum x\right)^2\right]\left[N\sum y^2 - \left(\sum y\right)^2\right]}}$$

You can probably see why I did not include it! If students want to try to use the formula, encourage them to research the meaning of Σ. It is a *summation* symbol that means to add copies of the expression immediately following it. For example, $\sum xy$ means to multiply each value of x by its corresponding value of y and then add all of the results, while $\sum x \sum y$ means to add all of the x values, add all of the y values, and then multiply the two results. N stands for the number of pairs (12 in this case).

Although some students may have fun learning about summation notation by playing with this formula and doing some calculations, it does not really help them understand the meaning of the formula. They will learn more about concepts underlying this formula when they take advanced mathematics courses. In addition to the many statistical ideas involved, r also has interesting connections to advanced algebra concepts such as vectors, inner products, and the law of cosines.

Exploration  9

Triangle Trials

The main goal of the Triangle Trials activity is to introduce students to a new type of probability. In all previous problems, it was possible (though not always easy!) to count the outcomes, because they were *discrete* (separate). In this exploration, the outcomes are *continuous*. That is, they "flow together" in a way that makes it impossible to list all of them!

For example, suppose you have a dartboard like this one.

Assuming that every point on the board is equally likely to be hit (and that you hit the board!), what is the probability of hitting a grey region? There are an infinite number of points that the dart can hit. It is not possible to use the formula:

$$P = \frac{\text{number of outcomes in the event}}{\text{number of outcomes in the sample space}}$$

because the numerator and denominator would both be infinity! Instead, calculating this probability involves finding the areas of the grey regions. (In fact, this would be an interesting problem for students to work out! The probability is 60%.)

The problems in Triangle Trials are carefully designed so that students begin with discrete probabilities that transform themselves into a continuous situation. It is likely that this transition will confuse students at first for exactly the reason described above; they will wonder how it is possible to find the value of "infinity over infinity."

However, if they have been drawing the suggested graphs along the way (please ensure that they do this!), and if they pay close attention to the visual patterns, they will be able to use area to make sense of the situation and to connect it to the discrete problems.

 DOI: 10.4324/9781003232780-12

STAGE 1

Suppose you have three line segments. Will they form a triangle? In Problem #1, students begin investigating this question from the perspective of probability. Before distributing the handout, I like to show students a picture of the subdivided line segment, read them the sentence below it, and ask what they notice and what they wonder. For example, they might observe that the segment is divided into 10 equal parts, that three segments will be formed after making the cuts, and that the longest possible segment will be 8 units. They may wonder what they are to do with the segments once they have them, or how many combinations of three segments are possible. Some of them may even suggest trying to make triangles and wonder if it is always possible to do this.

After discussing students' observations and questions, I share the handout with them, and talk about the directions. If the students came up with their own interesting questions during the introductory discussion, it is fun to follow up on these as well!

What You Will Need

» Graph paper

What Students Should Know

» Calculate probabilities for events containing equally likely outcomes.
» Graph ordered pairs in a coordinate grid.

What Students Will Learn

» Understand the conditions under which line segments will form a triangle.
» Apply knowledge of probability to solve a challenging problem.

Problem #1

Choose two of the small dots at random and imagine cutting the segment at those points.

Directions

- Determine the probability that the three resulting segments will form a triangle. Explain your thinking.
- Create a graph showing the pairs of cut-points that form a triangle and those that do not. Make some observations about your graph.

Testing the Waters

Answer the question for a line segment divided into fewer than 10 equal parts.

CONVERSATION STARTERS FOR #1

What do you notice? What do you wonder?

I notice that, in order to form a triangle, the two shorter segments must be able to "reach each other" when they are connected to each side of the longer segment.

I notice that it helps to keep a record of the lengths of the segments that are created by making the two cuts.

I notice the two cuts points are always different numbers.

I notice that once I list *a* and *b* as cut points, I should not list *b* and *a* as a separate outcome.

I wonder if there are formulas to predict these lengths from the cut points?

I notice that my list of cut points in the event (favorable outcomes) contains more numbers near the middle of the segment.

After students have created a graph:

I notice that the outcomes form triangles when I graph them.

I wonder why this happens?

I notice that counting the events involves finding sums of consecutive whole numbers.

I wonder if there is a quick way to find these sums?

Try using the commutative property of addition. (Reorder the numbers: first + last, second + second from last, etc.)

SOLUTIONS FOR #1

The probability is $\frac{6}{36}$ or $\frac{1}{6}$.

How to know if the segments will form a triangle: In order to form a triangle, the sum of the lengths of the two smaller sides must be greater than (not equal to) the length of the longest side. Otherwise, the shorter sides will not be able to reach to form the triangle! For example, you cannot make triangles from the segments in these two pictures.

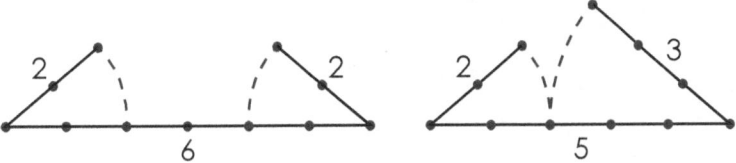

A strategy for calculating the probability: Begin by labeling the points on the segment.

List all pairs of cut points, and identify the lengths of the three line segments for each.

CP	SL	CP	SL	CP	SL	CP	SL
1,2	1,1,8	2,4	2,2,6	3,7	3,4,3	5,7	5,2,3
1,3	1,2,7	2,5	2,3,5	3,8	3,5,2	5,8	5,3,2
1,4	1,3,6	2,6	2,4,4	3,9	3,6,1	5,9	5,4,1
1,5	1,4,5	2,7	2,5,3	4,5	4,1,5	6,7	6,1,3
1,6	1,5,4	2,8	2,6,2	4,6	4,2,4	6,8	6,2,2
1,7	1,6,3	2,9	2,7,1	4,7	4,3,3	6,9	6,3,1
1,8	1,7,2	3,4	3,1,6	4,8	4,4,2	7,8	7,1,2
1,9	1,8,1	3,5	3,2,5	4,9	4,5,1	7,9	7,2,1
2,3	2,1,7	3,6	3,3,4	5,6	5,1,4	8,9	8,1,1

CP: cut points   SL: segment lengths

Six of the 36 equally likely outcomes are in the event (shaded), meaning that the probability is $\frac{6}{36}$.

A graph: The points in the sample space are dots and those in the event are the large dots.

Pairs of cut points that form triangles:

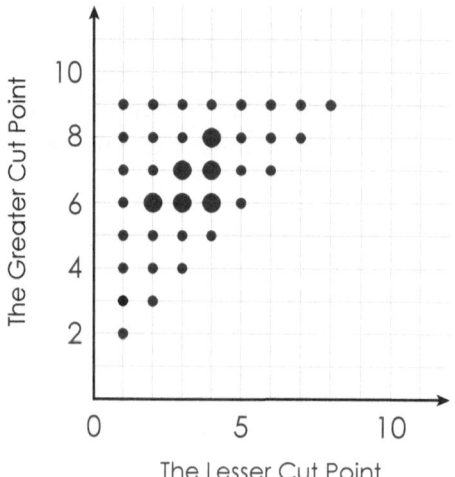

Some observations: The outcomes in the sample space form a right triangle, while those in the event form a smaller, inverted right triangle inside it that does not quite touch the sides of the larger triangle. Both triangles are symmetrical about the line through the points $(0, 10)$ and $(10, 0)$. The triangle for the events contains $1 + 2 + 3 = 6$ points, while the remaining dots in the sample space form three triangles, each containing $1 + 2 + 3 + 4 = 10$ dots. The number of outcomes in the sample space is $1 + 2 + 3 + 4 + 5 + 6 + 7 + 8 = 36$.

STAGE 2

One of the most important ways that mathematicians create new mathematical knowledge is through the process of *generalization*—extending concepts so that they apply in a broader context. In Problem #2, students experience the process of generalization when they subdivide the line segment from Problem #1 into more parts and then explore the results. They search for patterns in their numbers and graphs and use them to make predictions. By paying close attention to the details of their work, they may also discover new facts about the cut points that form triangles and learn more about why some pairs work while others do not.

The Diving Deeper question from Problem #2 challenges students to create an algebraic formula for the situation. The Algebra Connections pages at the end of the exploration have more information about this.

What You Will Need

» Graph paper

What Students Should Know

» Understand Problem #1.

What Students Will Learn

» Create and interpret graphs.
» Identify and extend patterns in order to generalize results from Problem #1.

Problem #2

These segments are divided into more parts than the segment in Problem #1.

Directions

- Assuming the same situation as in Problem #1, create a graph for each segment showing the outcomes in the sample space and in the event.
- Use your graphs to determine the probabilities that the segments form a triangle in each case. Explain your thinking.
- Describe any patterns that you see in your work and your answers.

Diving Deeper

- Create a formula to calculate the probability that three pieces will form a triangle when the segment is divided into *n* equal parts.
- Explore segments that are divided into an odd number of parts.

CONVERSATION STARTERS FOR #2

What do you notice? What do you wonder?

I notice that there will be more outcomes to keep track of than there were in Problem #1.

I wonder if the probability will be greater than, less than, or equal to the answer from Problem #1?

I wonder if the graphs will look similar to the graph in Problem #1?

I notice that when the three segments form a triangle, one of the cuts always seems to be to the left of the midpoint of the segment.

I notice that when the three segments form a triangle, none of them is ever longer than half of the original segment.

I notice that when the three pieces do not form a triangle, one of the pieces is always at least as long as half of the original segment.

I wonder if there is a formula for adding the whole numbers 1 through *n*?

There is. If you have time, encourage students to discover it! One version of the formula is $S = \dfrac{n \cdot (n+1)}{2}$.

After students have calculated the probabilities:

I notice that the probability seems to increase when *n* increases.

I notice patterns in the fractions, especially when I do not simplify them.

I wonder what will happen to the probability as *n* continues to increase?

I notice that the probability seems to increase more slowly as *n* becomes greater.

SOLUTIONS FOR #2

The graphs:
Pairs of cut points that form triangles:

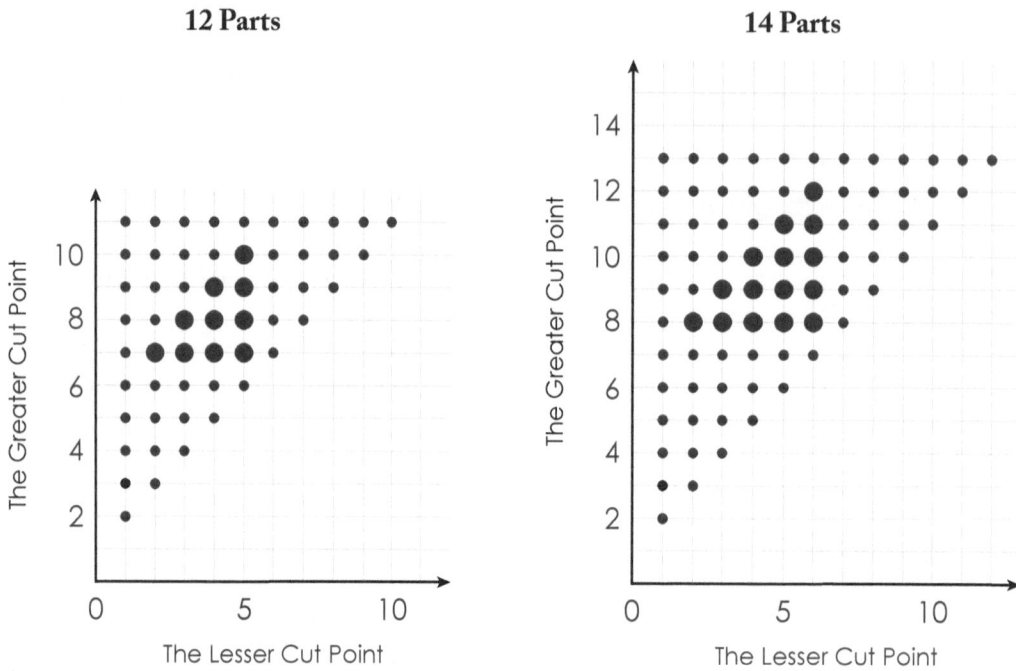

Some students may create these graphs merely by predicting them from the appearance of their graph in Problem #1. However, they should check that their cut points really do result in triangles!

The probabilities: For the 12-part segment: $\frac{10}{55}$ (or $\frac{2}{11}$). For the 14-part segment: $\frac{15}{78}$ (or $\frac{5}{26}$).

Some patterns: The general appearance of the graphs is the same as described in Problem #1. You may find the number of outcomes in each sample space using the same types of sums. For example, for the 12-part segment, there are $1 + 2 + 3 + 4 + 5 + 6 + 7 + 8 + 9 + 10 = 55$ outcomes in the sample space and $1 + 2 + 3 + 4 = 10$ outcomes in the event.

In general, if the original segment is divided into n equal parts, then the greatest number in each sum is $n - 2$ for the sample space and $\frac{n-4}{2}$ (or $\frac{n}{2} - 2$) for the event.

Other observations: Some students may notice that if there are n pieces, and the cut points are a and b (with a being the smaller one), then the lengths of the three

242

segments are a, $b - a$, and $n - b$. They may also observe that in order for these three segments to form a triangle, they must all be less than half the length of the original segment. In terms of the cut points, this means that a and $b - a$ must each be less than half of n, and b must be greater than half of n.

STAGE 3

After completing the first two stages, your students might wonder how it would affect the probability if they kept dividing the segment into smaller and smaller pieces until eventually they could make the cuts *anywhere* on the segment. This question takes them from the world of *discrete* probability, in which it is possible (though not necessarily practical!) to write down every outcome, into the world of *continuous* probability, where the number of outcomes is infinite. Of course, making lists and tree diagrams will no longer work. Fortunately, however, there is a new way to explore this type of situation—drawing pictures and calculating areas. In fact, the graphs that students created in Stages 1 and 2 have set the stage for them to do just this!

What You Will Need

> » Graph paper

What Students Should Know

> » Understand Problems #1 and #2.

What Students Will Learn

> » Use geometry to calculate probabilities in continuous situations.
> » Extend patterns to make predictions.

Problem #3

●━━━━━━━━━━━━━━━━━━━━━━━━━━━●

Imagine making random cuts at any two points along the segment.

Directions

- Predict the probability that the segments will form a triangle when you make the two cuts *anywhere* on the segment. (Use your results from Problems #1 and #2.)
- Draw a graph showing the pairs of cut-points for which the three segments will form a triangle. Is your graph consistent with your prediction? Explain.
- Test your graph by selecting points in the sample space that are inside (outside) the event space and checking that they form (do not form) triangles.

Diving Deeper

Perform an experiment in which you ask people to break a piece of spaghetti randomly in three pieces. Do you think that people actually choose the points randomly? How can you tell?

CONVERSATION STARTERS FOR #3

What do you notice? What do you wonder?

I notice that it is no longer practical to list all of the outcomes for the cut points!

I wonder how I should label the cut points now that there are an infinite number of them?

I wonder what the value of *n* is (see the Solutions to Problem #2) in this situation?

I wonder if I can predict the probability by extending the patterns in the fractions?

I wonder if I can predict the probability by extending the patterns in the graphs?

I wonder how the graph will change now that I can make the cuts anywhere?

I wonder if the graph will stay the same in some ways?

I notice that it helps to visualize a graph with a very large number of dots.

I notice that as the segment is divided into more parts, the vertices of the event triangle look like they get closer to the sample space triangle.

I notice that for outcomes in the event, the first cut point is always left of the midpoint.
   This may help you understand something about the graph.

I wonder if there is a formula for the probability for any *n*? (See the Diving Deeper question from Problem #2.)

I wonder what happens to this formula when you can make the cuts anywhere?

SOLUTIONS FOR #3

Prediction: Looking closely at the graphs from Problems #1 and #2, it appears that as *n* increases, the vertices of the "event triangle" are approaching the sides of the "sample-space triangle," and that the area of the event triangle may be getting closer to one fourth of the area of the sample-space triangle.

You may also write probabilities from the previous two problems as decimals:

$$\frac{6}{36} \approx 0.167 \qquad \frac{10}{55} \approx 0.182 \qquad \frac{15}{78} \approx 0.192$$

In fact, by using the sums described in the Solutions for Problem #2, you may even predict probabilities for segments with 16, 18, or 20 parts:

$$\frac{21}{105} = 0.2 \qquad \frac{28}{136} \approx 0.206 \qquad \frac{36}{171} \approx 0.211$$

The probabilities continue to increase, but more and more slowly, possibly approaching 0.25. Based on these observations, students may predict that the probability is $\frac{1}{4}$.

A sample graph:

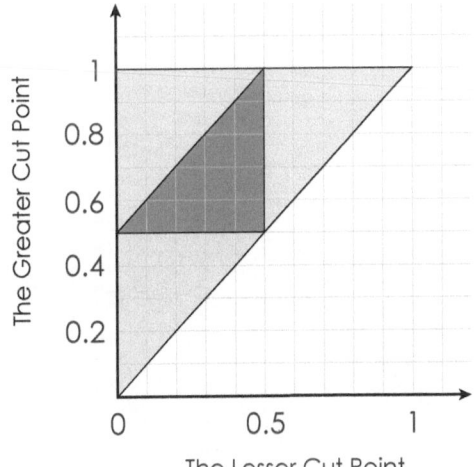

For this graph, I chose to set the length of the line segment at one unit. The inputs and outputs no longer have to be whole numbers, because they stand for *any* location on the segment where you make a cut. You might imagine the graph as an infinite number of points, infinitely close together. The event triangle (dark region) is one fourth the area of the larger triangle. Assuming that the graph is accurate, it is consistent with the prediction that the probability is $\frac{1}{4}$.

Testing the graph: (0.37, 0.52) is a point inside the event triangle. It corresponds to segment lengths of 0.37, 0.15, and 0.48. All three segments have lengths of less than 0.5 units, and the sum of the lengths of the two shorter segments (0.37 + 0.15 = 0.52) is greater than the length of the longest segment, 0.48. Therefore, these cut points will create a triangle as expected.

The point (0.23, 0.78) is not within the event triangle, although it is close. It corresponds to segment lengths of 0.23, 0.55, and 0.22. One of the segments is longer than 0.5 units, and the sum of the two shorter segment lengths is 0.23 + 0.22 = 0.45, which is less than the length of the longest segment. Therefore, these cut points will not create a triangle, again as expected.

By continuing to test points—especially points that are just barely inside or outside the boundary—students may provide more evidence that the graph is correct and that the probability is 0.25. The Algebra Connections on the next page show a couple of methods that further strengthen this evidence. By the way, notice that the points on the boundary of the event triangle do not belong to the event, but this does not affect the probability!

ALGEBRA CONNECTIONS

Students who have some experience graphing linear inequalities will be able to understand how the graph for Problem #3 comes directly from the conditions of the problem. If you set the length of the segment at 1, the graph of the sample space satisfies the inequalities:

$$x < 1, \; y < 1, \text{ and } y > x$$

(The final inequality follows from choosing x to be the smaller cut-point number.)

The graph is in the first quadrant, because all values are positive. The graphs of the regions are to the left of the vertical line $x = 1$, below the horizontal line $y = 1$, and above the diagonal line $y = x$, respectively. The three regions overlap in the large triangle that represents the sample space.

The graph of the event triangle is determined by the condition that all three segments must be less than half the length of the original segment. If you think about this, it means that the smaller cut point must occur to the left of 0.5 and the larger one to the right of 0.5. The difference of the cut-point numbers must also be less than 0.5. In symbols:

$$x < 0.5, \; y > 0.5, \text{ and } y - x < 0.5$$

These three conditions define the right, the bottom, and the diagonal side of the event triangle, respectively. Their graphs overlap to form the event triangle. Some students might like to think about how it would affect the inequalities and the graph if they chose x to be the larger cut-point number.

Students who are comfortable manipulating algebraic expressions may be able to create a formula for the probability that the three pieces form a triangle when the segment is divided into n equal parts. To begin, it helps to know a formula for adding the whole numbers 1 through n:

$$1 + 2 + 3 + \cdots + n = \frac{n(n+1)}{2}$$

Discovering this formula is an interesting challenge in itself! However, once students have it, they can go further. Looking at their graphs, they can see that the number of outcomes in the event is $1 + 2 + 3 + \cdots + \left(\dfrac{n}{2} - 2\right)$, and the number of outcomes in the sample space is $1 + 2 + 3 + \cdots + (n - 2)$. Using the formula above, they may challenge themselves to discover the expressions:

$$1 + 2 + 3 + \cdots + \left(\frac{n}{2} - 2\right) = \frac{\left(\frac{n}{2} - 2\right)\left(\frac{n}{2} - 1\right)}{2} \cdot \frac{4}{4} = \frac{(n-4)(n-2)}{8} \text{ and}$$

$$1 + 2 + 3 + \cdots + (n - 2) = \frac{(n-2)(n-1)}{2}$$

The probability is equal to the number of outcomes in the event divided by the number of outcomes in the sample space:

$$P = \frac{\dfrac{(n-4)(n-2)}{8}}{\dfrac{(n-2)(n-1)}{2}} = \frac{(n-4)(n-2)}{8} \cdot \frac{2}{(n-2)(n-1)} = \frac{n-4}{4(n-1)} \quad \text{or}$$

$$\frac{n-4}{4n-4}$$

Of course, students are often very creative, and they may come up with many other ways to discover this formula. At any rate, they can check it in at least a couple of ways. First, if they substitute $n = 10, 12, 14$, and so on, they obtain the same fractions (after simplifying) as in Problem #2: $\dfrac{1}{6}$, $\dfrac{2}{11}$, $\dfrac{5}{26}$, etc. Also, as they substitute larger and larger values for n, the fractions get closer and closer to $\dfrac{1}{4}$! This makes sense because, as n becomes greater, the "minus 4s" in the numerator and denominator matter less and less, and the value gets closer to $\dfrac{n}{4n}$, which equals $\dfrac{1}{4}$.

Appendix

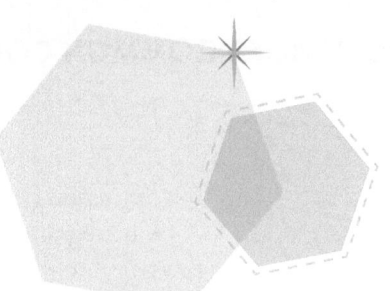

The Solutions in these explorations leave out the details for certain graphing and calculation procedures. Because some students may not yet have been introduced to these procedures, this appendix contains an example of each of the following:

» Creating a dot plot
» Creating a histogram
» Creating a box plot
» Calculating the interquartile range (IQR)
» Identifying outliers
» Calculating the mean absolute deviation (MAD)

Suppose that you have 22 heart rate measurements for sixth graders:

69, 71, 73, 79, 79, 80, 81, 81, 81, 81, 82, 84, 85, 87, 88, 88, 90, 91, 91, 96, 101, 111

Creating a dot plot (also known as a line plot):

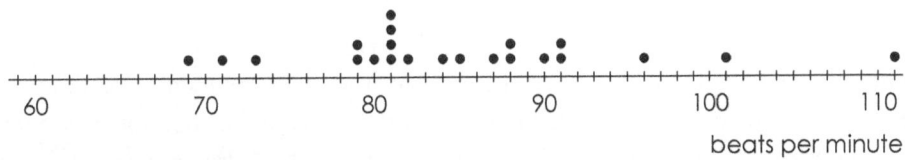

Place a dot on and/or above a number line for each data point. Align the dots neatly in columns and rows so that the height of each column clearly shows the number of occurrences. Be sure that the number line has a *uniform scale*.

On a uniform scale, equal differences between numbers always correspond to equal distances on the number line. For example, these two scales are *not* uniform:

Creating a histogram:

A histogram is a special kind of bar graph showing the number of data items (*frequency*) within a chosen interval. For example, suppose you choose an interval of 5 for the heart rate data. This interval (sometimes called the *bin width*) will be the width of each bar.

Heart Rates of Sixth Graders

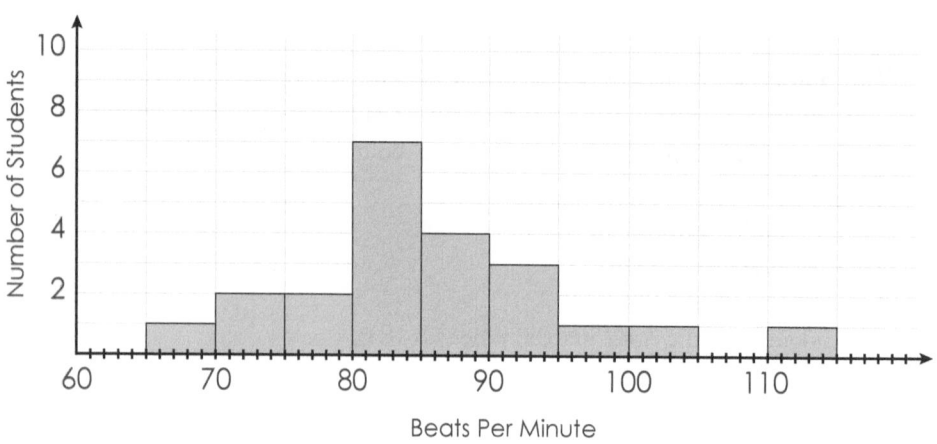

Beats Per Minute

This histogram shows that there is one student with a heart rate from 65–69, 2 with a heart rate from 70–74, 2 with a heart rate from 75–79, 7 with a heart rate from 80–84, etc. How would it affect the appearance of the histogram if you chose a larger or smaller interval? Try it!

Creating a box plot (also known as a box-and-whisker plot):

A box plot splits a set of data into four intervals, each of which contains one fourth of the data. Begin by finding the median. Then find the median of all numbers less than the median (the *lower quartile*) and the median of all numbers greater than the median (the *upper quartile*). From this, you get the *five-number summary* of the data:

minimum, lower quartile, median, upper quartile, maximum

For the heart rate example:

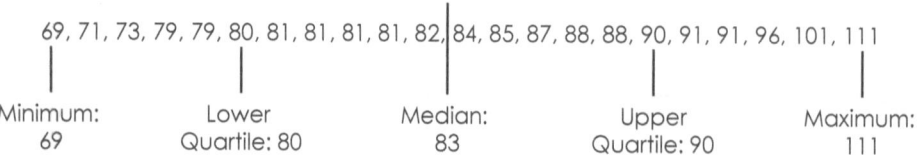

69, 71, 73, 79, 79, 80, 81, 81, 81, 81, 82, 84, 85, 87, 88, 88, 90, 91, 91, 96, 101, 111

| Minimum: 69 | Lower Quartile: 80 | Median: 83 | Upper Quartile: 90 | Maximum: 111 |

You can use these five numbers to create a *box plot*.

Heart Rates of Sixth Graders

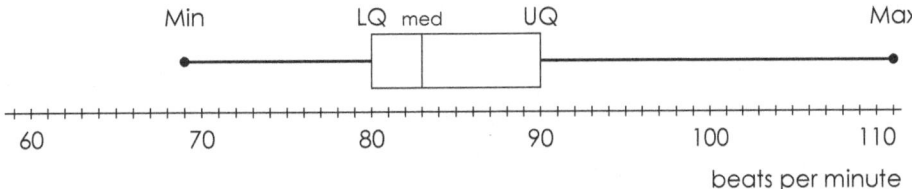

Minimum (Min): the left whisker
Lower Quartile (LQ): the left side of the box
Median (med): the bar inside the box
Upper Quartile (UQ): the right side of the box
Maximum (Max): the right whisker

Notice that in this example each of the four regions contains five data points (not counting the LQ and the UQ, because they are on the boundary between regions).

Calculating the interquartile range (IQR):
The *interquartile range* is simply the range of the middle half of a set of data. You can visualize it as the width of the box in the box plot. You calculate it simply by subtracting the lower quartile from the upper quartile. The IQR of the heart rate data is equal to 10 because:

$$IQR = UQ - LQ = 90 - 80 = 10$$

Determining if a data point is an outlier:
Roughly speaking, an *outlier* is a number that lies far outside the typical values of a set of data. More specifically, a number is an outlier if it is greater than 1.5 "box widths" (that is, 1.5 IQRs) distant from the box in the box plot.

For example, in the heart rate situation, the number 69 is not an outlier, because one and a half box widths is equal to the IQR · 1.5:

$$10 \cdot 1.5 = 15$$

and 69 is not greater than 15 units from the left side of the box (which is at 80). However, 111 is an outlier, because it is more than 15 units from the right side of the box (at 90). Consider challenging your students to create algebraic formulas (or inequalities) that determine whether a number is an outlier!

Calculating the mean absolute deviation (MAD):

The *mean absolute deviation* is the average of the distances of all of the numbers from the mean. The mean of the sixth-grade heart rates is about 84.95. To calculate the MAD, you begin by finding all of these distances:

$$|69 - 84.95| = 15.95$$

$$|71 - 84.95| = 13.95$$

$$|73 - 84.95| = 11.95$$

$$|79 - 84.95| = 5.95$$

$$|79 - 84.95| = 5.95$$

The sum of all 22 distances is 156.9. Therefore, the MAD is $\dfrac{156.9}{22} \approx 7.1$.

The absolute values simply ensure that the distance is always a positive number. In Exploration 1, students discover that if you find the *"mean deviation"* without using absolute values, the answer is always 0!

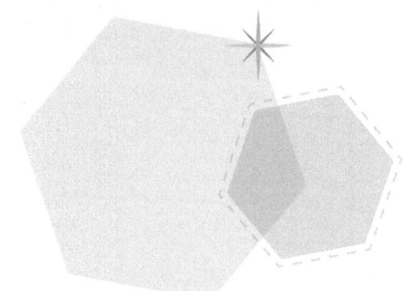

References

Alexie, S. (2007). *The absolutely true diary of a part-time Indian*. New York, NY: Little, Brown.

Chapin, S., O'Connor, C., & Anderson, N. (2013). *Classroom discussions in math: A teacher's guide for using talk moves to support the common core and more* (3rd ed.). Sausalito, CA: Math Solutions.

Dweck, C. (2007). *Mindset: The new psychology of success*. New York, NY: Ballantine.

Johnsen, S., Ryser, G., & Assouline, S. (2014). *A teacher's guide to using the Common Core State Standards with mathematically gifted and advanced learners*. Waco, TX: Prufrock Press.

Johnsen, S., & Sheffield, L. (2013). *Using the Common Core State Standards for mathematics with gifted and advanced learners*. Waco, TX: Prufrock Press.

Kilpatrick, J. (2001). *Adding it up: Helping children learn mathematics*. Washington, DC: National Academy Press.

National Governors Association Center for Best Practices, & Council of Chief State School Officers. (2010). *Common core state standards for mathematics*. Washington, DC: Authors.

Women's National Basketball Association. (2016). *Team stats*. Retrieved from http://www.wnba.com/stats/team-stats

Principles and standards for school mathematics. (2000). Reston, VA: National Council of Teachers of Mathematics.

Rowling, J. K. (1997). *Harry Potter and the sorcerer's stone*. New York, NY: A. A. Levine Books.

Sheffield, L. J. (2003). *Extending the challenge in mathematics: Developing mathematical promise in K–8 students*. Thousand Oaks, CA: Corwin Press.

Smith, M., & Stein, M. (2011). 5 practices for orchestrating productive mathematics discussions. Reston, VA: National Council of Teachers of Mathematics.

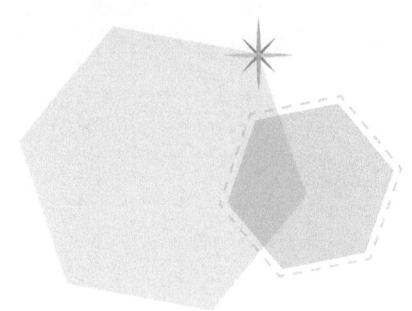

About the Author

Jerry Burkhart has been teaching and learning mathematics with gifted students for more than 20 years. He has degrees in physics, mathematics, and math education from University of Colorado, Boulder, and Minnesota State University, Mankato. Jerry provides professional development and presents regularly at conferences on meeting gifted students' needs in mathematics. He maintains a website at http://www.5280math.com.

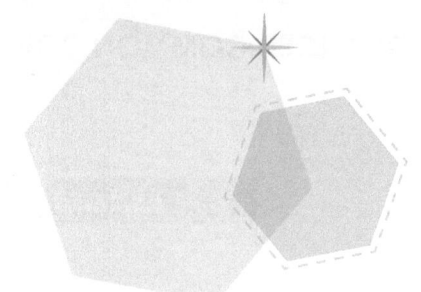

Common Core State Standards Alignment

Exploration	Common Core State Standards in Mathematics
Exploration 1: Playing With Data	6.SP.A Develop understanding of statistical variability. 6.SP.B Summarize and describe distributions.
Exploration 2: A Day at the Races	6.SP.A Develop understanding of statistical variability. 6.SP.B Summarize and describe distributions.
Exploration 3: Simulation Station	7.SP.C Investigate chance processes and develop, use, and evaluate probability models. 6.RP.A Understand ratio concepts and use ratio reasoning to solve problems.
Exploration 4: Comparing Populations	7.SP.A Use random sampling to draw inferences about a population. 7.SP.B Draw informal comparative inferences about two populations. 6.SP.A Develop understanding of statistical variability. 6.SP.B Summarize and describe distributions.
Exploration 5: One More Time!	7.SP.C Investigate chance processes and develop, use, and evaluate probability models. 6.SP.A Develop understanding of statistical variability. 6.SP.B Summarize and describe distributions.
Exploration 6: What Are the Chances?	7.SP.C Investigate chance processes and develop, use, and evaluate probability models. HSS-CP.B Use the rules of probability to compute probabilities of compound events.
Exploration 7: Paths and Pascal	HSS-CP.B Use the rules of probability to compute probabilities of compound events.

Exploration	Common Core State Standards in Mathematics
Exploration 8: Sports Correlations	8.SP.A Investigate patterns of association in bivariate data. 8.F.B Use functions to model relationships between quantities. 6.EE.C Represent and analyze quantitative relationships between dependent and independent variables. 6.RP.A Understand ratio concepts and use ratio reasoning to solve problems.
Exploration 9: Triangle Trials	HSS-CP.B Use the rules of probability to compute probabilities of compound events. HSA-REI.D Represent and solve equations and inequalities graphically.

ADVANCED
COMMON CORE
MATH
EXPLORATIONS